Framing a Drinking Water Supply Services Act for India, with Particular Reference to Rajasthan

Framing a Drinking Water Supply Services Act for India, with Particular Reference to Rajasthan

M. Dinesh Kumar

PETER LANG
Chennai - Berlin - Bruxelles - Lausanne - New York - Oxford

Bibliographic information published by the Deutsche Nationalbibliothek. The German National Library lists this publication in the German National Bibliography; detailed bibliographic data is available on the Internet at http://dnb.d-nb.de.

A catalogue record for this book is available from the British Library.

Library of Congress Cataloging-in-Publication Data

Names: Dinesh Kumar, M., author.
Title: Framing a drinking water supply services act for India, with particular reference to Rajasthan / M. Dinesh Kumar.
Description: Oxford ; New York : Peter Lang Publishing, 2024. | Includes bibliographical references and index.
Identifiers: LCCN 2024031377 (print) | LCCN 2024031378 (ebook) | ISBN 9781803746197 (paperback) | ISBN 9781803746203 (ebook) | ISBN 9781803746210 (epub)
Subjects: LCSH: Water-supply—Law and legislation—India—Rajasthan. | Water resources development—Law and legislation—India—Rajasthan. | Water rights—India—Rajasthan.
Classification: LCC KNS2522 .K86 2024 (print) | LCC KNS2522 (ebook) | DDC 343.5409/24—dc23/eng/20240716
LC record available at https://lccn.loc.gov/2024031377
LC ebook record available at https://lccn.loc.gov/2024031378

The funding for the original work on framing of the drinking water services Act for Rajasthan came from UNICEF-Jaipur. The author thankfully acknowledges their support.

Cover design by Peter Lang Group AG

ISBN 978-1-80374-619-7 (print)
ISBN 978-1-80374-620-3 (ePDF)
ISBN 978-1-80374-621-0 (ePub)
DOI 10.3726/b22075

Published by Peter Lang Pvt Ltd, Chennai
info@peterlang.com – www.peterlang.com

This publication has been peer reviewed.

Dedicated to my beloved Grandmother
Late Smt. Parvathi Pillayathiri Amma

Contents

Figures

Tables

Preface

Water is a unique substance. It is one of the few materials on Earth that exists naturally as a solid, liquid or gas. Life on Earth cannot exist without water. As a global phenomenon, the quality of water available in natural freshwater ecosystems – in rivers, phreatic aquifers and lakes – is deteriorating as a result of human activities, land use practices and economic development. Land use practices, such as the removal of forest cover, intensive cultivation of crops, livestock farming and aquaculture, affect the quality of water in our streams, lakes, groundwater and ultimately the marine environment.

According to the United Nations Development Programme (UNDP), water is 'the stuff of life and a basic human right' (UNDP, 2006). Thus, water is an essential element for life on Earth – including human life – and as a result, it is a core concern in law. From a legal perspective, the UNDP rightly emphasizes the importance of the human rights dimension of water. Yet, in practice, water law has to address the economic and environmental dimensions as well as the human rights dimension. In particular, historically, one of the central concerns of water law has been the development of principles concerning access to and control over water.

While water that is directly consumed, that is, the drinking water, is essential for the survival of human beings, it is also required for farming (growing crops, livestock rearing and fisheries), which produces food for human consumption, and manufacturing.

Despite the central role that water plays in sustaining lives and economies, the development of a formal water law has been relatively slow and often patchy. At the domestic level, colonial legislation first focused on the regulation of water for economic reasons, for instance, through the development of legislation concerning irrigation and navigation. Over the past few decades, increasing pollution of freshwater bodies, decreasing per capita water availability (resulting from increasing population) and excessive diversion of water for competitive uses have led to the development of water quality regulations and norms on environmental flows in rivers. All the water policies, from the first one in 1987 to the most recent in 2012, have prioritized drinking water in water allocation decisions.

In a civilized world, the 'right to a dignified life' cannot be guaranteed merely through the provision of an adequate volume of water to meet the physiological needs of the people. The 'right to live' should imply the right to food, water and a decent environment. Water is the key to achieving all these goals. The civil, political, social and cultural rights of people enshrined in the Universal Declaration of Human Rights and under the Constitution of India cannot be exercised without these basic human rights.

When water resources are scarce, their allocation and distribution need to be carefully regulated, given the growing competition between different water-use sectors. Framing a law on drinking water provisioning, though requiring small quantities of high-quality water, is therefore not an easy task under such circumstances.

Historically, irrigation laws constitute the most developed part of water law in India, though statutory water law also includes a number of pre- and post-independence enactments in various areas such as embankments, drinking water supply, irrigation and drainage, floods, water conservation, river water pollution, rehabilitation of evacuees and displaced persons, fisheries and ferries.

In India, the laws governing the use of water include international treaties and federal and state legislations. They also include a number of less formal arrangements, including water and water-related policies as well as customary rules and practices, which were weakened by formal central and state laws.

The drinking water laws of the Indian states largely deal with areas such as regulation of water supply by the agency, water charges, permit/license systems and strategy for water supply. The laws focus more on offences by the subscribers and are silent on the inefficiency in service delivery and the root causes. No Indian state has a comprehensive drinking water act that deals with the myriad of questions starting from the criteria for selection of water supply technologies to deciding on the service levels (quantity and quality of water, duration, etc.) to releasing the water connections to a new customer to charging for water to payment of penalties for non-compliance to disposal and treatment of wastewater to water supply surveillance.

This book, for the first time in the history of academic work on water laws in India, presents a draft act on drinking water services that comprehensively deals with the complex challenges of drinking water provisioning in a difficult landscape. It is done for one of the acutely water-scarce states of India, that is, Rajasthan, where water supply challenges are big and growing. Besides presenting the detailed act chapterwise with its seventy articles, it goes into the nuances of framing the act, starting from building the rationale for the specific provisions included. The book also presents a comprehensive review of various drinking water acts and other water legislations in India, and drinking water legislations in selected developed countries, viz., the United States, England and Wales, Japan, Canada, Malaysia, the EU, South Africa and two developing countries (Ghana and Bahamas), to evaluate and compare their unique features. The book also attempts to shed some light on the reasons for the failure or ineffectiveness of most water-related acts in India.

I sincerely hope the drinking water sector specialists, government policymakers in the water supply and sanitation sector and water law professionals in India and other developing countries will find the book useful for evaluating existing drinking water legislations and framing new ones.

Acknowledgements

First of all, I am extremely thankful to the two reviewers whose comments on the draft manuscript helped in enhancing the value of the final work. Thanks are particularly due to Ms Nandini Ganguli, Commissioning Editor, Peter Lang International Academic Publishers for her encouragement and immense support throughout the course of this work – starting from the conceptualization of the book proposal to the preparation and submission of the full manuscript.

I am highly indebted to Rushabh Hemani, my long-term associate from the UNICEF (United Nations International Children's Emergency Fund) field office in Jaipur, with whom I initially planned to write this monograph. Since 2016, he, on behalf of UNICEF-Jaipur, has been an integral part of the assignment along with the Water Supply and Sanitation Organisation (WSSO) of Rajasthan for drafting the 'Drinking Water Supply Services Act for Rajasthan', which formed the basis for writing this monograph. He heavily influenced several of the ideas and thoughts presented in this monograph. Due to certain administrative issues, he was asked to move out of this book-writing project, and he requested me to remove his name as the co-author of the monograph.

The project on drafting the above-mentioned Act actually began in 2015 with the support of UNICEF-Jaipur. Sincere thanks are also due to Shri Pankaj Mathur, who was then the WASH Specialist of UNICEF-Jaipur, for initiating the project in partnership with WSSO, Rajasthan.

I am also extremely grateful to my beloved daughter, Archana Dineshkumar, who currently works as the Consulting Junior Science Editor for IRAP, for meticulously editing the entire manuscript for language, coherence, consistency and style, and for checking and correcting the citations and references.

I owe profusely to my beloved family, especially my wife Shyma and my daughter who stood by me during the difficult times, provided moral strength and continue to motivate me to do more and more.

Lastly, I proudly dedicate this book to my beloved grandmother, Late Smt Parvathi Pillayathiri Amma whose love and affection towards me and her prayers continue to give me the courage to face the most difficult challenges of life. Like my late father, to a great extent, her ethos of life shaped my outlook on life too. She was a voracious reader and has been a great pillar of support for me during times of crises.

M. Dinesh Kumar
29 October 2023

Abbreviations

BIS	Bureau of Indian Standards
BOD	Biological Oxygen Demand
BPL	Below Poverty Line
CBOs	Community Based Organizations
CPCB	Central Pollution Control Board
CPHEEO	Central Public Health and Environmental Engineering Organisation
CrPC	Criminal Procedure Code
FBO	Food Business Operator
GPS	Global Positioning System
HRD	Human Resources Development
IEC	Information, Education and Communication
MNREGS	Mahatma Gandhi National Rural Employment Guarantee Scheme
NGOs	Non-Governmental Organization
NRDWP	National Rural Drinking Water Programme
O&M	Operation and Maintenance
PHED	Public Health Engineering Department
PRIs	Panchayati Raj Institutions
PWU	Public Water Utility
RWSSMB	Rajasthan Water Supply and Sewerage Management Board
SCADA	Supervisory Control and Data Acquisition

ToR	Terms of Reference
VWSC	Village Water and Sanitation Committee
WASMO	Water and Sanitation Management Organisation
WHO	World Health Organization
WQM	Water Quality Management
WSP	Water Safety Plan
WSSO	Water and Sanitation Support Organization

CHAPTER 1

Framing a Drinking Water Supply Services Act: Setting the Context

1. Introduction

There are several challenges in managing drinking water supplies of the required quality in adequate quantities and at adequate frequencies on a sustainable basis for protecting public health. They broadly concern finding a dependable source of freshwater or water that can be treated to potable standards with reasonable costs; fixing the criteria, norms and standards vis-à-vis the quantity and quality of the water to be supplied for drinking purpose; building the infrastructure for supplying this water to the communities; doing timely operation and maintenance (O&M) and finding technical and financial resources for the same; safe disposal of wastewater; hygiene education; prevention of water theft and contamination & pollution of source water; monitoring and surveillance of water supply systems to prevent contamination of water supplies; application of water supply legislations (WHO, 1997); and fixing a tariff for the supplied water that the users can pay and that can create sufficient revenues for the water supply utilities to manage their operations on a sustainable basis.

Such challenges become compounded when freshwater resources are scarce, financial resources to treat and supply water are limited, competition for water from other sectors of use such as irrigation and industry is high, water sources are highly susceptible to pollution and institutional capacity to plan, design, execute and manage water supply systems to ensure physical and environmental sustainability is limited.

Rajasthan is one of the most water-scarce regions in the world. Most parts of the state receive very low rainfall (100–400 mm) with high inter-annual variability and are characterized by high aridity (Kumar *et al.*, 2006;

Pisharoty, 1990). The annual per capita renewable water availability in the river basins of the state is one of the lowest in the country and in the world. There are very few pockets within the state that are reasonably well-endowed with water resources, and the rest of the state has very low to low and medium rainfalls and hyper-arid to semi-arid climatic conditions (Hemani *et al.*, 2021).

Historically, the communities in the region largely depended on traditional water harnessing bodies such as *nadis*, *tankas*, tanks and lakes, shallow open wells and *johad*s for meeting various water needs, particularly for drinking water supply and livestock drinking. These were by and large under the control of the community and were treated as community assets (Agarwal and Narain, 1997). Though they failed during prolonged droughts (Kumar and Pandit, 2020), for several millennia, these structures were able to meet the needs and expectations of the people (Agarwal and Narain, 1997). But, at that time, the population density of the region was very low; farming was subsistence-based with rain-fed crops and livestock rearing and extremely limited irrigation needs. Therefore, the pressure on water systems to cater to these needs was also low.

However, with changing socio-economic and cultural milieus and with the advent of modern water technologies and the government assuming greater responsibility for the provision of drinking water supply as a public duty, the expectations of the communities vis-à-vis the quantity of water to be supplied, the reliability of water supply and 'affordability' have gone up remarkably. Responding to these growing needs, the governments also plan to cover the maximum number of households in rural and urban areas with a formal water supply of 'tap water', that is, treated water that is free from physical, chemical and biological contamination and supplied through pipes. Drinking water supply is viewed as a 'public good', which should be available free of charge, by most people.

On the other hand, with the expansion of agriculture and growing preference for high-yielding, water-intensive crops such as wheat, cotton and chilli, the pressure on water resources for irrigation has become excessively high, even in the driest regions of the state. The natural catchments of rivers and streams, which provided fresh water inflows into ponds, tanks and lakes that once supplied water for cities and villages, are now heavily degraded

(Calder *et al.*, 2008). The local institutions, which once protected these traditional water bodies from pollution and other threats, no longer exist.

Groundwater resources are increasingly being exploited for commercial agriculture, with large parts of the state facing problems with aquifer mining (GOI, 2012; SWRPD, 2013). At the aggregate level, the total groundwater abstraction is 14,519 MCM per annum, against a renewable groundwater availability of 10,821 MCM (SWRPD, 2013). While groundwater resources are the major source of water for meeting domestic water needs in rural and urban areas of the state, depletion of groundwater caused by excessive dependence on it for irrigation has significant implications for the sustainability of water supply sources, with the drying up of public wells and hand pumps (Hemani *et al.*, 2021)[1]. The lack of well-defined rights and inefficient pricing of (canal) water and energy encourage their inefficient and unsustainable use (Kumar, 2010).

Extremely low population density in vast areas of the state, comprising the western, northern and north-western parts that are naturally water-scarce (Hemani *et al.*, 2021; Kumar *et al.*, 2008), poses new challenges, as modern, regional water supply systems that tend to achieve household water supply connections would be prohibitively expensive.

Along with water availability, there are serious concerns about the deteriorating quality of water in Rajasthan. The groundwater quality monitoring undertaken by the Public Health Engineering Department (PHED) of Rajasthan showed high levels of fluorides, nitrates, chlorides and TDS in 42,352 out of 75,266 observation wells. Groundwater contamination – with increasing levels of TDS, chlorides, nitrates, sulphates and fluorides associated with lowering water levels, tapping water from contaminated geological formations and pollution of groundwater and lakes from industrial

1 The absence of a sufficient number of large-scale surface irrigation facilities, geographically well-spread, combined with low-to-medium rainfall in most parts, high evapotranspiration demands for water, high frequency of droughts resulting from the departure of rainfall from mean values, and high per capita arable land (0.18 ha) increase the demand for irrigation water. Free power in agriculture continued for many years, and the existing pump horsepower-based pricing of electricity encouraged over-pumping and inefficient and often wasteful use of groundwater (Hemani *et al.*, 2021).

and urban effluents – is a major problem in many districts (SWRPD, 2013, p. 40)[2]. Not only are the existing water pollution control norms dated, but their enforcement is highly ineffective. This situation leaves the utilities with no option but to plan schemes that are based on the transfer of surface water over long distances. This involves huge capital and O&M costs. Adoption of such technologies also has negative implications for community involvement in their operation and maintenance.

As regards water supply services, there are no clear-cut 'user entitlements' for water for basic needs in volumetric terms, and ad hoc norms are still followed. For many decades, the supply norm followed by the state governments in India for rural water supply was 40 litres per capita per day, with no scientific basis. This norm was recently changed to make the per capita water supply in rural areas 55 litres per day (Kumar *et al.*, 2022). On the other hand, the cost of production and supply of water is not yet an important consideration in fixing water prices. Heavy subsidies for water provision threaten the financial viability of the schemes with high capital and O&M costs. In most instances, drinking water supply in rural and urban areas is neither metered nor priced on the basis of consumption, leaving little incentive for efficient use. Direct, illegal tapping of water from the public water supply systems is also becoming rampant, with serious consequences for access equity in water and revenues.

The demand for water in the domestic sector in the state is growing because of rising population, growing per capita water demands owing to rising incomes, changing lifestyles and demographic shifts. Growing water scarcity, increasing competition that the domestic water supply is facing from irrigation and industrial water supplies, a lack of community involvement in the management of the water supply at the local level, the resultant conflicts over water use between communities and the greater role the private sector is playing in the provision of drinking water through water markets, make it imperative that effective regulation is essential for a high-priority water use sector like the domestic water supply. Added to

2 This was also shown by the monitoring of the Central Pollution Control Board and State Pollution Control Board in Pali and Jodhpur districts to analyse the impact of wastewater on groundwater quality during 1994–2002.

these are the growing standards vis-à-vis quantity, reliability and quality of water supplied through public systems, changing notions of safe water, and growing demand for surveillance of water supply to prevent public health risks.

Such a regulation is essential for effective governance of the water supply sector in the state and improving water security. Water security is an important input for achieving progress in human development, and water insecurity can lead to several undesirable outcomes in terms of human development and social security[3]. As of 2007–2008, out of the twenty-three states for which data were compiled, Rajasthan stood seventeenth in terms of the human development index as per the HDI report.

A framework for decision making for new investments in the drinking water sector vis-à-vis the technologies to be chosen, the levels of investment costs and the service levels (per capita supply, the quality of water, frequency and duration, etc.), the price to be charged from the consumers, etc. is lacking in the state. Also lacking is the framework of actions to be taken in cases of applications for new water connections by residences and commercial establishments, non-payment of water charges, illegal water tapping, water pollution, non-compliance of regulatory measures, etc. that are legally tenable.

3 Lack of adequate water supplies can force people to compromise on water use for personal hygiene needs, at least for a short while, or use poor-quality water (Hutton and Chase, 2017). Poor adoption and use of modern sanitation systems in rural households is often linked to poor access to water in the dwelling premises (Wagner and Lanoix, 1959). Using open defecation as an alternative for personal hygiene poses a serious safety concern for women and girl children in rural areas (Padhi *et al.*, 2015). While drinking poor-quality water can cause waterborne diseases, prolonged exposure to poor-quality water for personal hygiene needs results in skin problems. If adults in households spend too much time fetching water, it can have significant opportunity costs for communities in terms of losing wages. Because of this, many school-going children are forced to find time to fetch water for domestic uses, resulting in poor school attendance. Private water markets, if allowed to flourish without any regulatory framework to control prices or ensure quality, can have long-term consequences for access to water for the poor.

2. Background and Rationale for Framing a Drinking Water Supply Services Act

Effective programmes to regulate drinking water supply depend ideally on the existence of adequate legislation, standards and codes. One of the functions of the basic legislation is to define the functions, authority and responsibilities of the water supply agency and the surveillance agency. Standards and codes should specify the quantity and quality of the water to be supplied to the consumer, the practices to be followed in selecting and developing water sources and in treatment processes and distribution systems, and procedures for approving water systems in terms of water quality. The precise nature of the legislation in each country will depend on national, constitutional and other considerations (source: based on WHO, 1997).

Rajasthan is a state that is severely drought-prone. An index map of the state is given in Figure 1.1. Every year, one or more parts of the state face drought. In spite of having government-run water supply schemes, hundreds of villages face acute drinking water shortages during droughts. One reason is that while groundwater pumping for watering crops intensifies on such occasions through deep wells, the domestic water schemes that depend on wells fail to get adequate quantities of water from underground during the summer months, just to even meet the basic needs of the rural population, as either the aquifers get fully depleted or the water levels drop alarmingly by the end of the winter season. Another reason is that in many situations, rampant construction of water harvesting structures such as check dams in the catchments of tanks and lakes that are used for water supply reduces the inflows into such water bodies. Pollution due to the disposal of untreated sewage by municipalities and large hotels was also reported in the case of some important lakes.

Agitations over the sharing of water between competing sectors (such as irrigation, domestic water supply and industrial water supply) from public reservoirs are also not rare in Rajasthan. In 2007, a large group of around 2,500 farmers protested the diversion of waters from Bisalpur dam to the city of Jaipur, located about 130 km away. These farmers from villages en

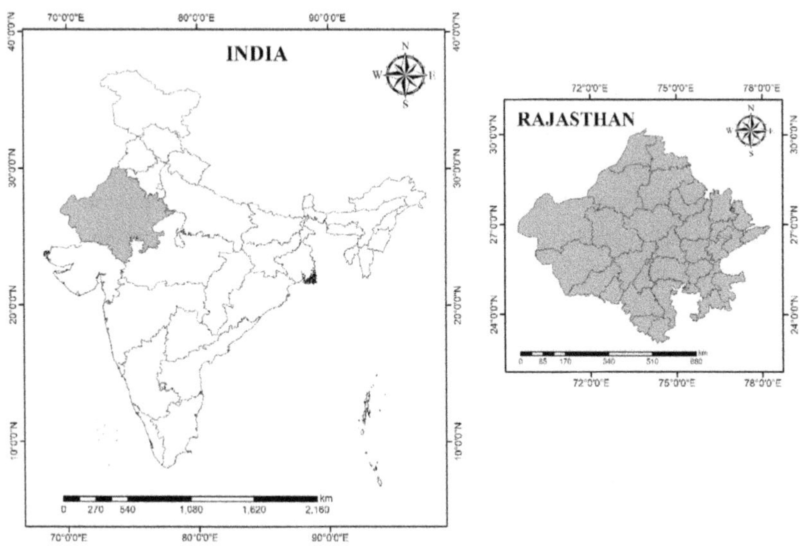

Figure 1.1: Index Map of Rajasthan in India

route to the pipeline that diverts the water from the reservoir to the city insisted that their drinking and irrigation water needs be met before catering to urban demands. The people had assembled at Soyla village in a gripping display of unity. Five farmers were killed when the police opened fire at the crowd. The agitation and the killings that followed forced the government to change the water allocation plan for Bisalpur dam to accommodate the demands of the protesting farmers (Bhaduri, 2015).

While there has been an overall increase in accountability on the part of the government agencies in the delivery of 'public goods', such as water supply, the provisions in the existing Acts are inadequate to tackle the emerging needs of theft prevention, water quality surveillance, penalties for non-compliance, legal remediation for recovery of water cess from defaulters, protection of sources from pollution and polluters, and overall institutional capacity building – comprising organisational change (private sector participation, involvement of community institutions), institutional reform and human resource development. There need to be clear guidelines on all of the above. The Act should clearly spell out what the potential role of the communities is in overall drinking water management and how the government will act in terms of involving the communities, PRIs and other local self-governments.

Also, clear-cut guidelines on the performance of the agencies in terms of treatment standards for raw water, wastewater treatment standards, water quality monitoring – comprising sampling and testing of water, collection, recording and reporting of information – staff efficiency[4] and the cost of production and supply of water are essential.

They also do not have the legal powers to regulate the functioning of various players in the sector to improve its performance and achieve equity, efficiency and sustainability. Illegal tapping of water distribution lines by households in cities and towns is rampant, causing not only revenue losses for the utilities but also service delivery problems for the tail-enders. Also, the operation of water tankers is unregulated. The administrative orders passed from time to time to protect the interests of the utility, and therefore the society at large, often do not stand the test of legal scrutiny in courts of law. Consequently, their effectiveness depends largely on how deftly the administration executes such orders.

In this context, a need was felt by the Water and Sanitation Management Organization (WSSO) of Rajasthan to frame a Drinking Water Supply Services Act for the state. The task of preparing the draft Act in collaboration with WSSO and the state office of UNICEF, Rajasthan, was given to the Institute for Resource Analysis and Policy (IRAP) in February 2015. The output of this assignment was to form the basis for framing a comprehensive Drinking Water Supply Services Act for the state. The assignment involved several consultations with domain experts from the region, including those who have an understanding of water law, senior officials in the Public Health Engineering Department (PHED) of the state of Rajasthan, consultants and community representatives, along with desk-top review and synthesis of various national and international guidelines pertaining to the quantity and quality of drinking water supply, water quality monitoring, water supply surveillance and water tariff fixing, private sector

4 The number of staff members per 1,000 households/population connected to the water supply system.

involvement in water supply and the existing water services and drinking water protection Acts in India and abroad.

The draft Act prepared by IRAP and UNICEF Rajasthan was presented to the senior and middle-level officials of the Public Health Engineering Department of Rajasthan[5] several times and to the Minister for PHED and Groundwater Development of the state once for comments and suggestions. The purpose was to ensure that the Act takes into account the realities on the ground vis-à-vis the socioeconomic and cultural background of the people and the institutional capacity of agencies that would be involved in implementing it. More importantly, it was important to make sure that none of the provisions of the Act drafted are in violation of the fundamental rights (of citizens) enshrined in the Constitution of India.

No Indian state has a comprehensive legal framework for the water supply sector that is capable of addressing the myriad of questions (Panickar, 2007)[6]. In the drinking water sector, such questions can range from the criteria for selection of water supply technologies to deciding on the service levels (quantity and quality of water, duration, etc.), releasing the water connections to a new customer, charging for water, paying penalties for non-compliance, disposing and treating wastewater, and water supply surveillance.

5 It included the Assistant Executive Engineers, the Executive Engineers, the Superintending Engineers, the Chief Engineer (Rural Water Supply) and the Principal Secretary, Public Health Engineering Department and Ground Water Department of Rajasthan.

6 Panickar (2007) points out that existing legislations largely deal with areas such as regulation of water supply by the agency, water charges, the permit/license system and strategy for water supply. Further, according to her, these water supply laws are characterized by limitations such as: water supply service being limited to only those who have obtained connections through formal applications, whether it is the household or industry; the state as the service provider regulates the supply and connections; charges are levied from the subscribers; individual households dependent on the connections are, for the most part, limited to towns and cities; the focus of the laws on offences by the subscribers, while the state authorities have no accountability; and the absence of any reference to the issues of service inefficiency and their causes (Panickar, 2007, p. 12).

More importantly, historically, irrigation laws constitute the most developed part of water law in India, though statutory water law also includes a number of pre- and post-independence enactments in various areas such as embankments, drinking water supply, irrigation and drainage, floods, water conservation, river water pollution, rehabilitation of evacuees and displaced persons, fisheries and ferries (Cullet and Gupta, 2009). Cullet (2016) notes that water law in Indian states is a patchwork of many state laws adopted over decades and some central legislation, the complexity arising out of which is made worse by the fact that there are many sectors that the water law has to address, and neither the lawmakers nor the institutions implementing them necessarily see the different sectors and their interlinkages. This is reflected in the separate treatment of surface and groundwater, both governed by different legal principles and addressed by different institutions that act independently. No country in the developing world has such drinking water Acts either. But drinking water supply challenges are enormous in these countries, with physical, economic and environmental sustainability challenges. Such an Act, if framed and enacted, can become a model for many other Indian states and even for other countries in South Asia and several countries in the African continent that lack basic access to water supply and sanitation to emulate. It will make decision making easy for the policymakers and planners in the drinking water sector.

3. Current Challenges in Public Water Supply in India

During the first five-year planning period (1951–1956), water supply and sanitation were added to the national agenda, and increased investments were made in the subsequent plans. The first National Water Policy (NWP) in India was adopted in 1987, which provided water allocation for drinking water as the first priority in the planning and operation of water systems. Subsequently, it was reviewed and revised in 2002 and then again in 2012 with a view to placing more emphasis on water conservation and promoting its equitable and efficient use. Further, for the first time, NWP 2012 treated water as an economic good and emphasized on ensuring access to a minimum quantity of potable water for essential health and hygiene for all citizens (NWP, 2012). States such as Andhra Pradesh, Assam, Goa, Himachal Pradesh, Karnataka, Kerala, Madhya Pradesh, Maharashtra, Orissa, Punjab, Rajasthan and Uttar Pradesh have already enacted and adopted their respective state water policy.

The National Water Policy (2012) recognizes that access to safe water for drinking and other domestic needs continues to be a problem in many areas. Public water supply is an essential service for communities and plays a vital role in general welfare, public health, drinking water security, economic activities and environmental preservation. However, the major challenges for achieving drinking water security through public water supply are: population growth; closing the coverage and service gap; ensuring sustainability of existing and new sources; and improving the quality of the services. In the water supply sector, it is very difficult to establish and maintain a really sustainable system. Nonetheless, the provision of safe water requires a service-oriented attitude on the part of water undertakings. Water services have to be normally set at an affordable level for the consumers and managed and operated in accordance with the regulations that are intended to protect the consumers and the environment.

Supply of safe drinking water in the water-scarce regions is one of the key responsibilities of any government. In India, the primary responsibility for providing drinking water facilities in the country rests with the respective state governments and, more specifically, with the local bodies in both rural and urban areas. Though various states have enacted regulations to ensure water supply and sanitation provisions, access to safe drinking water and sanitation still remains a dream for thousands of households. As per the UNICEF and WHO (2012) report on the progress of drinking water and sanitation, about 97 million people in India are without an improved drinking water source. For the supply of safe drinking water, an exceptionally high-quality institutional setup is indispensable. In water-scarce states like Rajasthan, institutions have to play a crucial role in drinking water supply management. Also, there is a need to reform the rules and regulations pertaining to the water supply.

In September 2019, the Ministry of Jal Shakti, Government of India, launched the ambitious programme called Jal Jeevan Mission (JJM) across the country for ensuring drinking water access to all through Functional Household Tap Connections (FHTC) by the end of 2024[7].

At the time of the launch of the mission in 2019, FHTC coverage in India was 16.72 per cent[8], and as of the end of the year 2022, the coverage had increased to 55.54 per cent. Meeting the objectives of covering the balance of 44.46 per cent households on a sustainable basis through the mission in the states is a major challenge considering the fact that it has a large base to cover with diverse terrains within a short timeframe of two years. The guidelines on source sustainability have been drafted in close alignment with the Jal Jeevan Mission to support and complement the on-going national flagship programme till 2024 and the sustainability of sources and systems beyond the timeline.

3.1. *Lack of a Water Security Framework*

The essence of water security is that concern for the resource base itself is coupled with concern surrounding the water services that exploit the resource base for human survival and well-being, as well as for agriculture and other economic production functions, and that the resource

7 The Mission focuses on community-based planning, implementation, operation and maintenance of drinking water schemes, implemented in a demand-driven approach led and sustained by the local self-governing institutions, to ensure water supplies at the household level in adequate quantity and quality. A Village Action Plan (VAP) is prepared by the Gram Panchayat or its sub-committee, that is, VWSC/Pani Samiti/User Group, etc., with support from ISA, PHED/RWS Department, DWSM based on a baseline survey, resource mapping and the felt needs of the village community. It also includes work on a water safety and security plan. This is transmitted under the District Action Plan, which undertakes the source sustainability measures required within the district and its costs. The State Action Plan, on the basis of the District Action Plan, plans for convergence for source sustainability, grey water management measures and O&M for the state. It promotes the management of water at the lowest appropriate level through decentralized, but integrated water resource management by the Gram Panchayats, with a key focus on water conservation, source sustainability and rainwater storage.

8 JJM Dashboard – Accessed on 20 April 2022.

should be developed and managed in an equitable, efficient and integrated manner. Achieving water security thus requires cooperation between different types of water users and between those sharing river basins and aquifers within a framework that allows for the protection of vital ecosystems from pollution and other threats. Water security is also a precondition of any effective poverty reduction strategy and of effective environmental sanitation, wastewater management and flood control (GWP, 2000).

The water crisis is often a crisis of governance, resulting from the failure to integrate policies and practices related to the management of water resources. Good water governance exists where government bodies responsible for water establish an effective policy and legal framework to allocate and manage water in ways responsive to national social and economic needs and to the long-term sustainability of the resource base (GWP, 2000).

While the National Water Policy (2012) and the water policies of many Indian states prescribe giving top priority to drinking water in water allocation decisions, there are no norms, rules or institutional mechanisms available for ensuring this as the approach (Kumar, 2018) to water resource development is highly sectoral. The water resources departments of Indian states are concerned with developing surface water resources of river basins primarily for irrigation and flood control and do not consider different water needs of the population (drinking & domestic uses, manufacturing, livestock watering and water for fisheries) and environmental flow requirements of the river downstream in a holistic fashion while planning large river valley projects (Kumar, 2010).

The state water supply agencies, which have limited financial resources and lack the technical capacity to plan and build reservoirs for water supply, have to depend on the water resources departments of the respective states if they have to secure rights to use large quantities of surface water for supply through multi-village piped water supply schemes. Therefore, they prefer groundwater-based schemes such as open wells, bore wells and tube wells that can be managed with low capital investments. During droughts, if the wells go dry, water from public reservoirs is earmarked for drinking water supply by the Collector of the concerned district as a 'knee-jerk' reaction to acute water shortages. Comprehensive, long-term planning of water resources at the river basin scale, considering the multiple water needs of the population and nature, is absent (Kumar, 2018).

As regards groundwater, the resource is largely in the private domain, with landowners having the right to use the groundwater underlying his/her piece of land under the English Common Law, which is still being followed in the country (Cullet, 2012; Kumar, 2018). While the state and central government agencies carry out resource evaluation and planning for groundwater, the volumetric allocation of groundwater for the drinking water sector, done for each district, is only on paper. While several millions of farmers tap groundwater using different types of wells from the extensive and localised aquifers, there is no institutional mechanism to effectively regulate groundwater withdrawal by these farmers, which is essential to ensuring that the water that is required to meet the drinking water needs during the summer months is left aside in the aquifer. Even monitoring of groundwater withdrawal by various users is absent. As a result, in many regions with limited groundwater potential, the drinking water wells go dry during the summer months.

3.2. Water Supply Challenges in Rajasthan

Groundwater is the major source of water supply in the state. Around 90 per cent of drinking water demand and 73 per cent of water demand for irrigation are met by groundwater. As a result, groundwater development has risen from 35 per cent in 1984 to 138 per cent in 2008. The groundwater development is expected to reach 189.1, 221.4 and 259.1 per cent by 2015, 2020 and 2025, respectively, thus making it a severely overexploited region. In sixteen of the thirty-three districts, groundwater is overexploited (groundwater development of more than 100 per cent), and the rate of exploitation is as high as 165 per cent in Jhunjhunu and 153 per cent in Jodhpur. Since a small fraction of the total arable land is currently irrigated, groundwater exploitation will only worsen in the future as the area under irrigation is likely to expand. This would have a direct impact on water supply schemes that tap underground sources.

Two-thirds of the state is the Great Thar Desert, which suffers from recurrent water scarcity. With only one perennial river, the Chambal, the state has no dependable surface water. Famine conditions often prevail due to scanty rainfall. There are fourteen defined river basins and an outside basin in Rajasthan. Available surface water is confined mainly to the south and southeast parts of the state. A large area in the west does not have a defined drainage basin. Thus, the state's water resources are not only disproportionately low when compared to its people, livestock and farmers, but are also distributed unevenly.

On the basis of a more realistic indication of available surface water (50 per cent dependability), the state has about 16.05 BCM of economically usable surface water. The state has created capacity to harness and store 11.29 BCM, or around 70 per cent of available water. An irrigation potential of 3.4 million hectares has been created, against a potential of 5.1 million hectares. Rajasthan has around 11.36 BCM of renewable groundwater. Dependent on inflows into the rivers, 17.88 BCM is allocated through inter-state agreements, although it is not dependable due to political compulsions of the upper riparian states (Hooda, 2013). The summary of water resource availability in the state is presented in Table 1.1.

The data presented in Table 1.1 needs to be interpreted very carefully. The economically utilisable surface water resources, which are apparently higher than the usage, are estimated at 50 per cent dependability. This

Table 1.1: Water Resource Summary for Rajasthan

Category	Availability in BCM	Usage in BCM	Percentage Utilized
Economically usable internal surface water	16.05	11.29	70
Usable ground water	11.36	11.77	104
Usable interstate / external water	17.88	12.66	71
Total economically usable water in the state	45.29	35.72	79

Modified from source: Hooda 2013

means in 50 per cent of the years, the water availability will be less, and in several of those years, it can be just half of what is available at 50 per cent dependability (i.e. 16.05 BCM). In those years, the reservoirs built for storing the water will not be full, and the usage can suffer heavily. In the hilly southeastern region, groundwater is available only in some parts of the valley as the water table fluctuates rapidly due to the rocky geographical formations in the area. Reservoirs in the area dry up and create severe water crises during years of continuous drought.

The western desert area and Aravalli region often face drought conditions due to scanty and erratic rainfall. Further, there are only a few reservoirs in these areas. There were good aquifers in the region. But increased withdrawal of water for agriculture, industry and domestic use is fast depleting groundwater and affecting its quality. This is evident from the water accounting study carried out in the Luni River basin. The study shows that out of the total consumptive water use of 2741 MCM per annum during 2011–2012, a total of 2404 MCM (88 per cent) was used in irrigation, while domestic water use was only 4 per cent. Further, the total annual consumptive use exceeded the renewable water resource of the basin, which was estimated to be 2604 MCM for that year (Kumar and Bassi, 2013). These analyses reinforce the fact that the drinking water scarcity experienced in Rajasthan is necessarily an allocation problem and not a problem of physical availability *per se*.

The available groundwater has high TDS, fluoride and nitrates in most parts, especially in the alluvial areas in the western parts of the Aravalli ranges.

The Ganga and Bhakra canal systems, the Indira Gandhi Nahar Pariyojna (IGNP) in the northwest and the Narmada canal system in the south are the only reliable drinking water sources in the region. But the water supply schemes built around these sources are regional schemes, catering to several villages and sometimes towns and cities. They pose new management challenges. The high population growth rate (one of the highest in the country at 21.44 per cent decadal growth), along with its geographical problems, and the distribution over sparsely populated pockets, have aggravated the problem of providing safe drinking water. The population density in the state has increased from 129 per sq. km in 1991 to 165 per sq. km in 2001 and 201 per sq. km by 2011. Around 70 per cent of habitations in the state face problems getting potable drinking water.

During the eleventh five-year plan (2007–2012), twenty-four major water supply projects amounting to USD 404.5 million were completed in the state. Further, USD 2.67 billion was allocated for water supply in the twelfth five-year plan (2012–2017). Out of this, USD 372.8 million was proposed for urban and rural water supply schemes in 2013–2014. These schemes include twenty-one major water supply schemes and nine new projects to help meet the growing drinking water needs of the state.

4. Objectives of the 'Drinking Water Services Act'

The goal of drafting the act was to evolve a set of standards, criteria and guidelines pertaining to all aspects of water supply provision in rural and urban areas of Rajasthan to be enforced by a statutory body, covering the line agencies in the water supply, private water suppliers and civil society in their ambit. They would form the basis for drafting a 'Water Services Act' for the state of Rajasthan, which would then be debated and passed in the legislative assembly. The objectives of the study were as follows:

1. Define 'safe water', 'risk free water' and 'rights to access water' in the context of domestic water supplies and 'sources of drinking water'.
2. Evolve clear-cut standards for the treatment of raw water for public water supply and the treatment and disposal of wastewater.
3. Evolve norms and standards regarding the quantity, frequency and quality of water to be supplied through 'drinking water sources' managed by public utilities.
4. Define criteria and evolve norms for fixing water tariffs in rural and urban areas.
5. Define criteria for the selection of water supply technologies.
6. Frame guidelines for the location, design, construction and operation of water supply and wastewater treatment systems.

7. Evolve protocols for surveillance of the water supply for the prevention of public health hazards and risk prevention.
8. Frame guidelines on the functioning of private water providers and drinking water markets in rural and urban areas.
9. Frame guidelines pertaining to 'illegal water tapping' and source protection and define criteria for fixing penalties for non-payment of water charges, illegal water tapping and contamination of drinking water sources.
10. Evolve criteria for evaluating the performance of water supply agencies, and frame norms on water quality monitoring and sampling & testing of water.
11. Evolve a framework for institutional capacity building in the water supply sector, comprising institutional reforms, organisational strengthening and human resource development, including local institutional development and private sector participation.

5. Approach and Methodology for Drafting the Act

The methodology used in the study would be eclectic. It is discussed below. The approach would consist of desktop reviews to synthesize the state-of-the-art knowledge in the field, expert inputs and stakeholder consultations.

1. *Review of existing guidelines from international agencies (WHO, UNICEF, UNDP, etc.) and Indian government agencies* pertaining to quantity and quality of drinking water, 'water safety', risk prevention, surveillance of water supply, water quality monitoring, pricing of water, private sector involvement in water supply, and various enactments.
2. *Review and synthesis of studies on water supply and health, water surveillance, performance of water supply systems*, particularly the factors influencing the performance of water supply systems in both rural and urban areas, and the factors that increase the

vulnerability of communities to problems associated with poor quality water supply. There is a large volume of international literature on the themes of the performance of water supply systems, the linkage between water supply and human health and the factors influencing the vulnerability of communities to the health consequences of poor water supply.

3. *Analysis of existing guidelines for the preparation of legislation pertaining to the drinking water sector*: The guidelines for the preparation of legislation in the drinking water sector, prepared by the IELRC (International Environmental Law Research Centre, New Delhi), are studied so as to avoid conflicting laws and to evolve acts that are socially viable, legally tenable and implementable.
4. *Review of water service acts in the states of India and other countries*: Several countries have regulatory frameworks for the water supply sector, including drinking water protection acts. They include the various regulatory frameworks concerning drinking water, wastewater and drinking water sources that exist in the state of British Columbia in Canada, the Safe Drinking Water Act (1974–2004), the Water Services Act (2007) of the Republic of South Africa, the EU Water Services Act 2007, the EU (Drinking Water) Regulations 2014, the Waterworks Act 1957 of Japan, etc. From our review, we found that twenty-six Indian states and a UT have acts pertaining to either drinking water provision, sewerage or wastewater treatment, protection of water supply sources, or some of it or all of it, by the states, the Municipal Corporations and the Municipalities[9]. They are being reviewed to understand the key design features. The existing rules in the PHED of the state of Rajasthan would be reviewed and adaptations made.

9 They are: Andhra Pradesh, Arunachal Pradesh, Assam, Bihar, Chhattisgarh, Delhi, Goa, Gujarat, Haryana, Himachal Pradesh, Jharkhand, Karnataka, Kerala, Maharashtra, Madhya Pradesh, Manipur, Mizoram, Nagaland, Odisha, Punjab, Rajasthan, Tamil Nadu, Tripura, Uttarakhand, Uttar Pradesh, West Bengal and the UT of Andaman and Nicobar.

5. *Consultation with regional experts on water supply and sanitation*: Periodic consultations will be held with water and sanitation experts from the region, especially those who have worked on legal and regulatory aspects, on the nuances of framing acts pertaining to water supply. The idea was to understand the key provisions of such an Act in the context of Rajasthan and the crucial factors that need to be taken into consideration while drafting such acts to prevent loopholes and ensure success in implementation.
6. *Review of water policy and water administration in the drinking water sector in Rajasthan*: The state of Rajasthan had a water policy in 2010. The document would be reviewed for key policy objectives and the instruments for ensuring drinking water security in the state. Also, the organisational structure, rules and regulations, administrative norms and procedures, staffing and human resource capabilities of the state water supply administrative apparatus would be reviewed for necessary modifications or strengthening.
7. *Review of experiences with regulatory measures in the water sector from other regions*: There are many provincial Acts in the water resources and water supply sector that have been in force for quite some time. For instance, the Maharashtra Water Resources Regulatory Authority Act (MWRRA), which deals with basin-wide inter-sectoral water allocation, bulk water pricing and the water entitlement of user groups and users within the basin, has been in force since 2007. The impacts of these regulatory measures would be reviewed in order to gain insights into the factors that are crucial for success or failure in the enforcement of these measures. There are regulatory instruments in the groundwater sector to check over-exploitation, used by institutional financing agencies such as NABARD and State Electricity Utilities in several states, such as well spacing regulations and denial of electricity connections for agro-wells, respectively. Their effectiveness was also studied through reviews of impact assessment studies and other scholarly works.

6. Purpose and Scope of the Monograph

The purpose of this monograph is to discuss the process of drafting a Drinking Water Services Act for a province or state under a federal system of governance by presenting what goes into identifying the various provisions or essential ingredients of the Act, and building the rationale for keeping those provisions vis-à-vis the criteria, norms, rules, guidelines and standards pertaining to various aspects of the drinking water sector, and then to finally present the Drinking Water Services Act drafted for the state of Rajasthan in western India. The basic aim is to familiarize the water sector professionals working on the legal and regulatory aspects of water-related services such as drinking water supply with the nuances of drafting a law. The Rajasthan case is used to explain how various data on the physical environment, socioeconomic characteristics and institutional set-up can be used to evolve various provisions of the Act, especially the criteria, norms, rules, guidelines and standards pertaining to various aspects of the drinking water sector.

Having discussed why such an Act is essential for a state like Rajasthan in this introductory chapter, the monograph analytically presents background information about the drinking water scenario in the state of Rajasthan in sufficient detail, covering the characteristics of the resource, the water supply technologies that exist, the types of access to water supplies by the communities, the water pricing scenario, the regulations that exist to control water theft and illegal tapping of water, wastewater treatment and reuse, policies with regard to private sector involvement and water supply service providers, issues about the surveillance of water supply, and institutional capacity in the drinking water sector in Chapter 2.

In Chapter 4, the monograph then presents the draft drinking water Act framed for the state of Rajasthan, covering the eighteen chapters and the seventy articles it has, line by line. Prior to presenting the contents of the Act, a detailed review of various Acts of select municipalities, corporations and Gram Panchayats of fifteen Indian states that have bearing on water supply and sewerage as well as the drinking water and sanitation Acts of a few Indian states were reviewed and discussed vis-à-vis the frontiers they covered and their strengths and weaknesses in Chapter 3. The concluding chapter (Chapter 5) investigates why many water-related Acts in India fail or become ineffective.

CHAPTER 2

Building the Rationale for Framing a Drinking Water Services Act for Rajasthan

1. Introduction

According to the Global Water Partnership, water services refer to the supply of water for domestic or industrial needs and for sanitation. It includes the abstraction, treatment and distribution of water as well as the collection, treatment and disposal of sewage through a sewerage system or direct on-site sanitation. The provisions for legislating and regulating water services include the structure, ownership and control of the industry, the specification of duties to supply and standards of supply, economic regulation of tariffs, customer services and water efficiency. The regulation must consider both social and environmental goals, along with economic goals. Additional aspects include governance arrangements, both within the service providers and in relation to their regulation, to ensure transparency, accountability and participation. Water legislation may seek to address equity issues by mandating provisions for providing water necessary to satisfy basic human needs (GWP, 2017).

In this chapter, we seek to explore the various provisions that a drinking water services act for Rajasthan should have vis-à-vis the criteria, norms, rules, guidelines and standards pertaining to various aspects of the drinking water sector. These aspects include the provision of water connections; water supply (quantity, quality and frequency); wastewater treatment; wastewater recycling and reuse; water pricing; rainwater harvesting; water supply surveillance; protection of drinking water sources; operation and maintenance (O&M) of water supply systems; regulation of groundwater development; and institutional capacity building in the drinking water sector. But transposing the laws made elsewhere in the world, or for that matter, strictly following the guidelines or framework laws developed for the country as a whole without understanding the climatic, environmental, socio-economic, cultural, and political contexts in which they were made will not help either in enforcement or in bringing out the desired outcomes, as all those factors will have an influence on the appropriateness of certain legislations and rules in a given context.

Therefore, the chapter discusses the characteristics of water resources in Rajasthan, the current water supply challenges posed by resource availability and quality problems and technological and institutional constraints, the characteristics of water supply schemes in the state and access to water supply by households, adequacy and safety of water supplied for domestic uses, especially the quantity, quality and frequency of supply, pricing of water in the domestic sector, issues of private sector involvement in water supply and regulations of service providers, rules and regulations pertaining to 'illegal water tapping', the various considerations for choosing the water supply technologies, the various physical, social and economic dimensions of rainwater harvesting, and wastewater treatment and reuse. It also discusses some of the emerging concepts, such as water supply surveillance and capacity building in the water supply sector, relevant for the state of Rajasthan.

2. Water Resources in Rajasthan

Rajasthan has a very low per capita renewable water availability of 600 m^3 per annum (Jat *et al.*, 2009), which is far below the national average of around 1,500 m^3 per annum. More importantly, there is notable spatial variation in renewable water availability. The internal renewable water resource availability in the state is governed by rainfall, climate, topography, soils and aquifer characteristics. All of these variables display a high degree of spatial heterogeneity. The sub-sections below describe how they vary across the state.

2.1. Rainfall, Climate, Soils and Physiography

There is remarkable spatial variation in rainfall in Rajasthan. The highest average mean annual rainfall is in Banswara and the lowest in Jaisalmer. The mean annual rainfall in the state varies from 250 mm in the northwestern parts in Jaisalmer to 1,100 mm in the southeast. The coefficient of variation in rainfall is also very high, particularly in the lower rainfall regions, with values as high as 60 per cent. The number of rainy days decreases gradually from thirty-one to forty days in the southeast to less than twenty days in the northwest (Pisharoty, 1990).

The Aravalli ranges, lying in the NE-SW direction, have a marked influence on the rainfall in Rajasthan. There is a sharp reduction in the amount of rainfall on the western side of the Aravalli ranges, making western Rajasthan the most arid part of India. The average annual rainfall is as high as 1,278.5 mm in Bhilwara.

The climatic variable that poses the greatest water management challenge to the state is high evaporation. The annual reference evapotranspiration values range from 1500 mm in the southern part of the state to 2,000 mm in the western part in Jaisalmer (GOI, 1990). The annual potential evaporation is also very high, as evident from the map. The spatial trend in reference evapotranspiration in the state is almost opposite to the spatial trend in rainfall. However, in some parts of Udaipur, the reference evapotranspiration is as low as 1,400 mm.

The different types of soils found in Rajasthan are desert soils with sand dunes extending over the entire western and northern Rajasthan; red desertic soils found in the south central parts covering most parts of Jodhpur, Pali and Jalore districts; red loam found in parts of Udaipur, the entire Dungarpur and parts of Banswara and Kota districts; old alluvium found in most parts of Jaipur, Ajmer, Alwar and Tonk districts; recent alluvium found in the entire eastern side of Rajasthan covering most parts of Dholpur, Sawai Madhopur, Bundi and Kota; yellowish brown soils in parts of Bhilwara, Chittorgarh and Udaipur; and Sierozeme found in parts of Nagaur and Jaipur districts. In addition, red and yellow soils are found in the foothills of the Aravalli ranges.

Soil infiltration rates show wide variation across Rajasthan, from 0.63 cm/hour to 32/5 cm/hour. The infiltration rates are extremely high in the desert soils found in Churu, Bikaner, Jaisalmer and Ganganagar (2.04–32.5 cm/hour) and in the old alluvium found in Tonk, Alwar and Jaipur (0.82–21.0 cm/hour). They are relatively lower in the recent alluvium found in parts of Chittorgarh, Sawai Madhopur, Bundi, Udaipur, Dholpur and Bharatpur (0.63–12.0 cm/hour) and in the black soils found in parts of Jhalawar, Kota and Banswara (0.86–2.77 cm/hour). Infiltration is high in the red desertic soils found in most parts of Jodhpur, Jalore, Pali and small parts of Jhunjhunu and Sikar (11.82 cm/hour measured only at one location). Sierozeme found in parts of Jaipur, Nagaur and Pali has infiltration values ranging from 8.11 to 25.2 cm/hour.

Physiographically, Rajasthan can be divided into seven distinct parts, viz., western sandy plains (0–150 m) having 80–100 per cent sand dunes; Ghaggar plain with 40–80 per cent sand dunes (150–300 m); sandy arid plains with slight to 40 per cent sand dunes; semi-arid transitional plains with slight to 40 per cent sand dunes but at an elevation of 300–450 m; Aravalli range and hilly regions from 600–900 m and above; northeastern hilly region; eastern plains; Vindhyan scarp land and plateau; and Deccan lava plateau. Both the western sandy plains and the sandy arid plains have almost no drainage. The semi-arid transitional plain is drained by the Sukri River and the Luni River. The Aravalli ranges in the southern part are drained by rivers flowing towards the south, viz., Banas and Sabarmati. The eastern plains, the Vindhyan scarp land and plateau and the Deccan lava plateau are drained by the Banas River and Chambal River, flowing in the eastern and northeastern directions, respectively.

2.2. *Groundwater Resources and Heterogeneity in Resource Availability and Quality*

Figure 2.1 shows the geohydrological map of Rajasthan. The state of Rajasthan has heterogeneous groundwater conditions – unconsolidated, semi-consolidated and fully consolidated formations, with varying groundwater potential.

The unconsolidated formations are porous formations. The aquifer potential varies widely between 40–100 litres per second (lps) for the very good ones, 10–40 litres per second for moderately good ones, and less than 10 lps for low potential ones. The semi-consolidated formations include: clay-stone, sandstone, grit, siltstone, conglomerate and limestone. They also form porous aquifers and have groundwater potential varying from less than 10 lps to 100 lps. The consolidated formations are fissured rocks. The yield of the aquifers varies widely between 5 and 10 litres per second (good for fissured rocks) and below one litre per second. Hilly aquifers are found in very small pockets in south Rajasthan.

Unconsolidated porous regionally extensive formations with low groundwater potential cover the northern, northeastern and southern parts of the state, comprising the entire Ganganagar and most parts of Churu, Pali, Balmer, Jaipur and Alwar districts. But there are some pockets within these unconsolidated formations that have high yields. These patches are in Jhunjhunu, Sikar and Sirohi districts. But the groundwater underlying the entire Ganganagar and Churu districts is saline and therefore unfit for irrigation and drinking. Unconsolidated discontinuous aquifers with poor yield potential cover the western parts of the state, including the Thar Desert in Jaisalmer. Most of it is saline, except for some patches in the extreme west.

Semi-consolidated aquifers of low yield potential are found in parts of Jaisalmer, Bikaner and Churu districts. There are patches of semi-consolidated aquifers with moderate yield potential in Balmer and Sirohi districts and those with high yield potential in the lower northwestern parts, covering parts of Bikaner and Churu districts. These aquifers are inherently saline. However, over the years, the quality of groundwater in this region has improved, the reasons for which will be discussed in the next section.

Consolidated fissured formations of sandstone and shale with low yield are found in Nagaur and Jodhpur districts, and those with moderate yield potential are found in other parts of Nagaur, Jodhpur and Jaisalmer districts. Consolidated limestone and dolomite formations in small patches are found in Nagaur, Jodhpur and Jaisalmer districts.

Consolidated fissured formations of meta-sedimentary and meta-volcanic origin with low yield potential (1–5 lps) are found in the southern parts of Rajasthan, extending up to the central part of the state. They cover the entire Udaipur and Dungarpur districts and parts of Bhilwara, Jalore, Chittorgarh and Tonk districts, as well as some pockets in Jaipur and Alwar districts. Consolidated fissured sedimentaries with low yield potential are found in parts of Jalore, Bhilwara, Chittorgarh and the entire Sawai Madhopur, Dholpur and Kota districts. Basalts with inter-trappean clay are found in the southern part of Jhalawar and the eastern part of Dungarpur and Banswara districts.

The total amount of renewable groundwater resources in the state is estimated to be 10,791.8 MCM per annum, and most of it is concentrated in the alluvial and sedimentary formations (SWRPD, 2013). In addition to these dynamic groundwater resources, the alluvial and sedimentary formations in western Rajasthan also have large amounts of static groundwater resources, which make the region susceptible to mining.

One major factor that affects the availability of groundwater for domestic uses is the poor quality of groundwater in many areas. There are four major natural contaminants in groundwater in Rajasthan, viz., salinity, chlorides, fluorides and nitrates, whose concentrations exceed the permissible limits for drinking water. The incidence of groundwater quality problems due to these contaminants is more widespread in the western side of the Aravalli ranges. In many areas in the Thar Desert, the salinity levels in groundwater even exceed permissible levels for irrigation (source: based on Hemani *et al.*, 2021).

2.3. Surface Water Resources in Rajasthan

Rajasthan does not have a great history of surface irrigation, unlike south India. Though the state has several traditional water harnessing systems such as the *johads*, *khadins*, *nadi*, *kund* and tanks, their contribution to expanding irrigation hasn't been remarkable.

The state has very few major rivers, namely, Sabarmati, Banas, West Banas, Banganga, Chambal and Mahi. Banas is part of the Chambal River system. The total Chambal catchment is 72,032 sq. km (Table 2.1). Only the upper catchments of these major rivers (Mahi, Banas, West Banas,

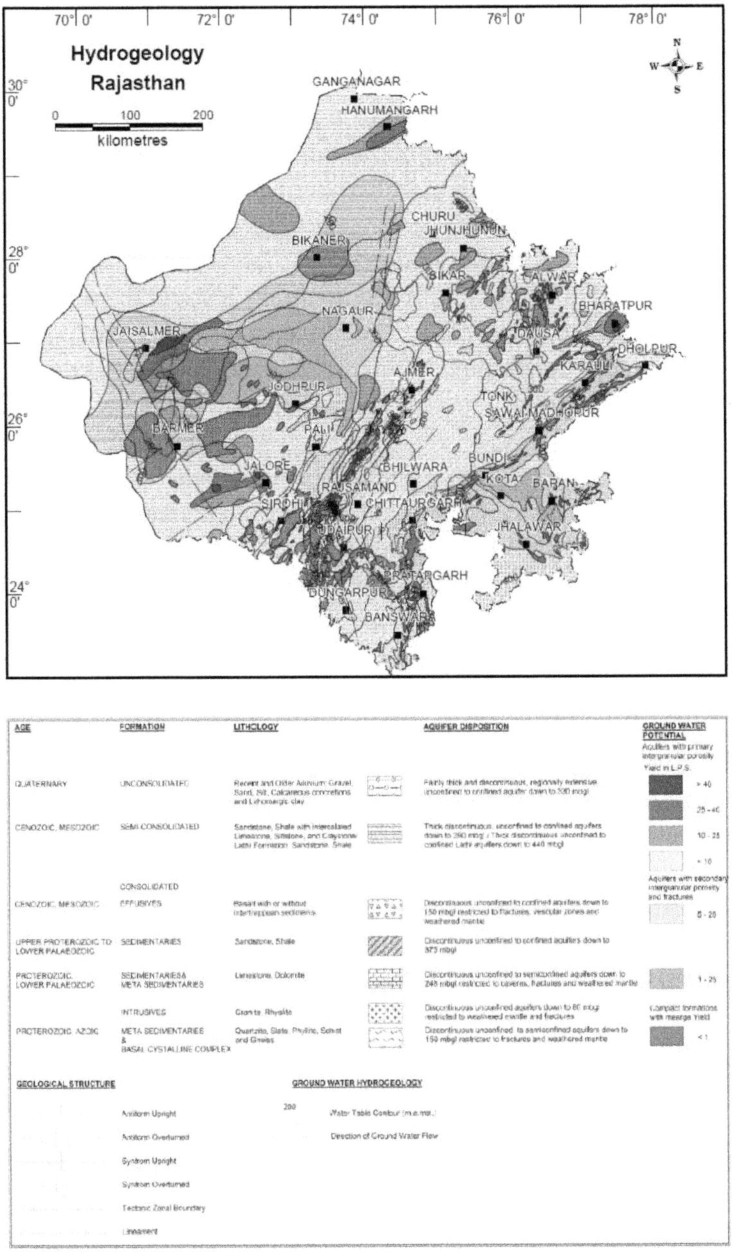

Figure 2.1: Geohydrological Map of Rajasthan

Sabarmati and Chambal) are located within the state. As a result, the state's rights to water from these basins are extremely limited. All the rivers originating from the state, including the major rivers, are seasonal in nature. Some of the west-flowing rivers of the state are highly ephemeral in nature and carry stream flows for very few days during the rainy season. In western Rajasthan, most of the area's drainage (except that of the Luni River) is internal, and streams are lost in the desert.

Rajasthan has several large freshwater lakes. Some of them were major sources of water for some cities. The Rajsamand Lake, Fateh Sagar Lake and Pichola Lake in Udaipur; Ana Sagar Lake and Pushkar Lake in Ajmer; and Gadisar Lake in Jaisalmer are the most important freshwater lakes in the state.

The total surface resource potential of Rajasthan (at 75 per cent dependability) is estimated to be 14,122.20 MCM per annum and that with 50 per cent dependability is around 21,713 MCM per annum. Of this, the total runoff that can be economically tapped is around 10,448 MCM. Most of the surface water resources are, however, concentrated in the south and southeastern parts of the state, and the western parts generate too little runoff. However, most river basins in Rajasthan are already over-appropriated, with the storage capacity of the existing reservoirs in these basins exceeding the utilizable runoff. There are only two river basins in Rajasthan that have some untapped runoff. They are the Chambal basin and the west Banas basin, in addition to the non-basin area. They account for a total of nearly 710 MCM of water (SWRPD, 2013).

The state also receives a total of around 17,880 MCM of water annually from the Yamuna (1,120 MCM), Rabi-Beas (11,970 MCM), Sutlej (1,740 MCM) and Narmada (620 MCM) rivers.

2.4. *Total Renewable Water Resources in Rajasthan*

The estimates of renewable water resources in Rajasthan available from the report of the State Water Resources Planning Department are presented in Figure 2.2 in graphical form. On average, the per capita renewable water availability in the state is merely 532 m^3 per annum. The

Table 2.1: Description of River Basins in Rajasthan

Name of the Basin	Description	Drainage Area (sq. km)	Amount of Utilizable Surface Water (MCM) for Future Use
Ahar River	Ahar River is a tributary of the Berach River. It originates in the hills of Udaipur District and flows through the city of Udaipur before joining the Berach.	3,516.50*	
Chambal River	The Chambal River remains one of North India's most unpolluted rivers, home to a rich diversity of flora and fauna. The National Chambal Wildlife Sanctuary is famous for the rare Ganges River dolphin.	18,446.45	572.70
Banas River	The Banas is a river in Rajasthan. It is a tributary of the Chambal River, which in turn flows into the Yamuna, a tributary of the Ganges. The Banas is approximately 512 km in length.	33,760.05	
Banganga River	River Banganga originates in the Aravalli Hills, near Arnasar and Bairath in Jaipur District. It flows towards the south up to the village of Ghat, then east through partly hilly and partly plain terrain.	6,742.57	

(*continued*)

Table 2.1: Continued

Name of the Basin	Description	Drainage Area (sq. km)	Amount of Utilizable Surface Water (MCM) for Future Use
Sabarmati River	The Sabarmati River Basin is situated in the mid-southern part of Rajasthan, between latitudes 23°25' and 24°55' and longitudes 73°00' and 73°48'.	3,288.68	
Luni River	The Luni River Basin is located in south-western Rajasthan, between latitudes 23°41' and 27°05' and longitudes 71°04' and 74°42'.	34,866.4	
Mahi River	The Mahi is a river in western India. It rises in Madhya Pradesh and, after flowing through the Vagad region of Rajasthan, enters Gujarat and falls into the sea through a wide estuary near Cambay. Its total length is 500 km, and its drainage area is 40,000 sq. km.	16,551.18	
Gomati River	The Gomati River is a small river. It originates in the hills of central Udaipur District, flowing south to join the Som River in the southern part of the district. The river was dammed in the seventeenth century to create Dhebar Lake, also known as Jaisamand Lake, which has an area of 50 km^2.	Tributary of Mahi River	

Table 2.1: Continued

Name of the Basin	Description	Drainage Area (sq. km)	Amount of Utilizable Surface Water (MCM) for Future Use
West Banas River	It originates in the southern Aravalli ranges in the Sirohi District of the state of Rajasthan. It flows south, draining the valley between Mount Abu on the west and the easterly ridge of the Aravallis on the east. It continues south through the plains of the Gujarat state, flowing through Banaskantha and Patan districts, and finally empties into the seasonal wetland of the Little Rann of Kachchh	2,837.81	138.10

*Note: The drainage area is of Barah basin.

estimates show that in all but one river basin, the per capita renewable water resources in the state are far below the threshold of 1,000 m^3 per annum, and in many river basins it is even below 500 m^3 per annum, making them 'absolutely water scarce'. The basins that fall under the latter category are: Shekhawati, Luni, Banganga, Banas and Ruparal. The basins that are water scarce are: Gambhir, West Banas, Gaghar, Mahi, Sukhi, Parbati and Sabarmati. Chambal is the only basin with per capita renewable water resource availability exceeding 1,000 m^3 per annum.

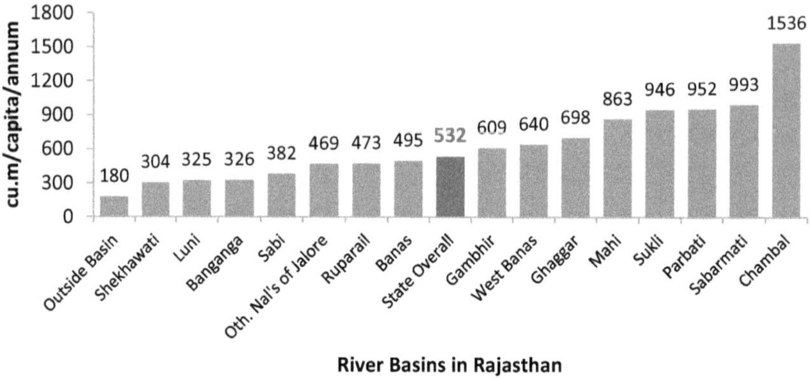

Figure 2.2: Average Renewable Water Availability in Rajasthan

3. Water Supply Schemes and Access to Drinking Water Sources in Rajasthan

In the state of Rajasthan, the Public Health Engineering Department (PHED) is responsible for providing drinking water supply to rural and urban areas. The types of water supply schemes that exist today in the state and the degree of access that communities have to different types of 'water supply sources' would probably indicate whether there is any real need for a drinking water act for the state or not.

As Table 2.2 indicates, there are five different types of water supply schemes in rural Rajasthan. They are: (1) pipes, pumps and tank schemes for individual villages; (2) hand pump-based schemes; (3) regional water supply schemes, which serve several villages; (4) TSS; and (5) Diggies and other schemes. Among these, the most common type is hand pump-based schemes, which serve nearly 48 per cent of the villages in the state. Individual village schemes, which involve pumping devices, pipes for water distribution and an overhead tank for water storage, cover nearly 12.5 per cent of the villages in the state. Such schemes are mostly groundwater-based schemes. The second most common type of scheme is regional water supply

Table 2.2: Number of Different Types of Water Supply Schemes in Rural Rajasthan

District	Piped, Pump and Tank Schemes	Hand Pumps	Regional Water Supply Schemes	TSS	Diggies and Others	Total
Ajmer	95	524	358	13	35	1,025
Alwar	270	1,389	110	38	147	1,954
Banswara	47	1,264	32	32	0	1,375
Baran	120	949	20	0	0	1,089
Barmer	339	0	1,567	27	0	1,933
Bharatpur	262	735	367	0	0	1,364
Bhilwara	253	1,091	280	36	33	1,693
Bikaner	355	9	279	23	138	804
Bundi	71	730	38	0	0	839
Chittorgarh	148	1,378	26	0	0	1,552
Churu	126	0	655	27	46	854
Dausa	111	805	109	0	0	1,025
Dholpur	51	694	41	0	0	786
Dungarpur	97	648	98	11	0	854
Ganganagar	71	177	2,322	0	260	2,830
Hanumangarh	175	420	1,014	0	164	1,773
Jaipur	628	974	273	2	200	2,077
Jaisalmer	86	105	396	0	13	600
Jalore	114	16	567	0	0	697
Jhalawar	135	858	484	0	0	1,477
Jhunjhunu	118	38	35	665	0	856
Jodhpur	474	9	566	9	0	1,058
Karauli	201	445	109	0	0	755
Kota	60	566	186	0	0	812
Nagaur	572	48	559	63	238	1,480
Pali	168	236	352	29	151	936

(*continued*)

Table 2.2: Continued

District	Piped, Pump and Tank Schemes	Hand Pumps	Regional Water Supply Schemes	TSS	Diggies and Others	Total
Pratapgarh	50	835	26	0	0	911
Rajsamand	140	764	69	0	0	973
Sawai Madhopur	116	531	72	0	0	719
Sikar	101	204	24	656	0	985
Sirohi	65	223	85	70	12	455
Tonk	57	677	292	5	1	1,032
Udaipur	202	1,872	99	5	0	2,178
Total	5,878	19,214	11,510	1,711	1,438	39,751

Source: Progress Report-2011–2012, PHED, Rajasthan

schemes, which are based on either surface reservoirs, river lifting or river bed French wells. They cover nearly 28 per cent of the villages in the state.

The physical characteristics of the water supply systems would, to a great extent, define the nature of the institutions that are needed to manage them (Kumar, 2014). In the case of hand pumps, which cover nearly half of the villages, direct involvement of the state water supply and sanitation department in the post-construction management is absent. So is the case with the individual village-based schemes. The PRIs will have to play a key role in the management of water supply for effectiveness. The performance of these decentralized schemes would depend heavily on the capacity of the local institutions, such as the village water and sanitation committees, that are created to maintain them regularly. Whereas, in the case of regional water supply schemes, which involve more sophisticated management, higher order institutions might also be required for management of the 'main system' along with the village institutions, which would handle water distribution at the village level. The effectiveness of these institutions would depend on their legitimacy, that is, how far they are recognized by law and what kind of institutional capabilities they have.

Table 2.3 shows the data on the types of 'water supply sources' to which the individual households have access. This is different from the 'types of schemes', discussed in the earlier paragraphs. The source also implies types of resources (water body) being tapped (groundwater, lakes, ponds, springs, etc.), the quality of the water and the reliability and dependability that can be expected. Here it is evident that though 48 per cent of the villages have hand pump facilities for water supply (Table 2.1), only 31 per cent actually use water from these sources (see Table 2.2). Others might be depending on wells/tube wells, bore wells or tanks/ponds that are in the private domain. This indicates that hand pumps are not reliable sources of water. The poor reliability could be because of seasonal depletion of groundwater if these hand pumps are located in hard rock areas.

Nearly 27 per cent of the households depend on 'tap water'. As per the definition of the Central Public Health Engineering Research Institute (CPHERI), this is treated water supplied through pipes and available in the dwelling premises through a tap. Hence, ideally, this would be water coming from surface reservoirs/lakes or from canals or river lifts, and not groundwater, as generally groundwater is not treated since it is a decentralized source. However, this doesn't mean that all the water that is supplied from rivers/canals is treated. As Table 2.6 indicates, nearly 1.1 per cent of the dwellings access water from rivers and canals, without treating. From the table, it is evident that nearly 70 per cent of the households depend on sources that cannot guarantee water that is safe from physical, chemical and biological points of view. Within these 70 per cent, in instances where the communities depend on water sources that are 'exposed' (ponds, tanks, lakes, rivers and canals), there could be high chances for biochemical and bacteriological contamination. In the case of underground sources (hand pumps, tube wells/bore wells/wells), the source could be potentially contaminated with physical and chemical contaminants, given the fact that groundwater contamination with fluorides, nitrates and high salinity is rampant in Rajasthan.

As regards urban areas, nearly 83 per cent of the households had access to tap water, and the majority of the remaining 13 per cent depended on hand pumps (6.1 per cent), bore wells and tube wells (5.6 per cent). The above data and analysis lead us to the point that there is a need for better

Table 2.3: Main Sources of Drinking Water (of Individual Households)

Source of Water Supply	Rural	% in total	Urban	% in total	Total	% in total
Tap Water	25,54,095	26.9	25,51,782	82.6	51,05,877	40.6
Well	13,14,715	13.9	46,975	1.5	13,61,690	10.8
Hand pump	29,88,588	31.5	1,89,499	6.1	31,78,087	25.3
Tube well/ Bore hole	13,65,468	14.4	1,72,418	5.6	15,37,886	12.2
Spring	7,982	0.1	1,231	0	9,213	0.1
River/Canal	1,02,026	1.1	3,612	0.1	1,05,638	0.8
Tank/Pond/ Lake	6,87,267	7.2	50,321	1.6	73,75,88	5.9
Other Sources	4,70,222	5	75,102	2.4	5,45,324	4.3
Total	94,90,363	100.1	30,90,940	99.9	1,25,81,303	100

Source: Census, 2011

clarity on which category of sources could be termed 'safe water' sources on the basis of the existing or new norms on drinking water quality, a legal framework that ensures access to 'safe water' for the communities and grievance redressal in case that access to safe water is denied.

4. 'Adequate' and 'Safe Water' Supply

As per Article 21 of the Indian Constitution, drinking water has been recognized as a basic human right. All human beings are entitled to an equal and non-discriminatory supply of a sufficient amount of safe water. To provide access to safe water supply and adequate sanitation for poor people, on 28 July 2010, the UN General Assembly recognized that safe and clean drinking water and sanitation are human rights essential to the full enjoyment of life and all other human rights. Subsequently, at its fifteenth session in September 2010, the UN Human Rights Council affirmed that the right to water and sanitation is derived from the right to an adequate standard of living and is inextricably related to the right to the highest attainable standard of physical and mental health, as well as to the right to life and human dignity. The combined effect of the two resolutions was to anchor the right to water and sanitation in the framework of the right to an adequate standard of living, making it legally binding like any other of the rights inscribed in UN treaties.

4.1. Quantity, Quality and Frequency of Supplied Water

4.1.1. Quantity of Water for Domestic Water Supplies

The importance of adequate water quantity for human health has been recognized for many years, and there has been extensive debate about the relative importance of water quantity, water quality, sanitation and hygiene in protecting and improving health. Despite this debate, international guidelines or norms for minimum water quantities that domestic water supplies should provide remain largely lacking. For instance, the Millennium Declaration (Development) Goals include a target to 'halve the proportion of people who are unable to reach or afford safe drinking water by 2015', but they do not specify in what quantity such water should be supplied. The WHO/UNICEF Joint Monitoring Programme, which produces the Global Assessment of Water Supply and Sanitation data, describes reasonable access as being 'the availability of at least 20 litres per person per day from a source within one kilometre of the users' dwelling' (WHO and UNICEF, 2000). However, it should be noted that this definition relates primarily to access and should not necessarily be taken as evidence that 20 litres per capita per day is a recommended quantity of water for domestic use.

Norms for quantities of water to be supplied have been proposed for certain specific conditions. For instance, the SPHERE project sets out 15 litres of water used per capita per day as a key indicator of meeting minimum standards for disaster relief (SPHERE, 1998). In the guidance manual prepared for the Department for International Development (UK), Well

(1998) suggested that a minimum criterion for water supply should be 20 litres per capita per day, while noting the importance of reducing distance and encouraging household connection. A similar figure has been suggested by other researchers (for instance, Carter *et al.*, 1997). Gleick (1996) suggested that the international community adopt a figure of 50 litres per capita per day as a basic water requirement for domestic water supply. It is important to distinguish between quantities of water required for domestic purposes (which primarily influence health and productivity) and quantities of water required for other purposes (such as agriculture, industry, commerce, transport, energy and recreation). Overall, the requirements for domestic supply typically constitute a very minor component of total water withdrawals (Gleick, 1993; 1996).

The PHED of Rajasthan had adopted norms on the quantity of water to be supplied for domestic purposes in rural and urban areas, as given in Table 2.4.

The per capita water supply to be maintained by government schemes as per the notification of the Chief Engineer (HQ), PHED, Rajasthan Jaipur, vide his letter no. CE(R)/2012–13/4008 dated 13 August 2012, under the new guidelines of the NRDWP issued by MORD, Govt. of India, is also given in the last column of Table 2.5.

Table 2.4: Norms on Per Capita Water Supply per Day as per PHED, Rajasthan

Category of Areas	Per Capita Water Supply	Per Capita Water Supply (as per revised norms)
Rural		
Drought-prone areas	70 litres per day	100 litres per day
Non-drought-prone areas	40 litres per day	70 litres per day
Urban		
Towns having population more than 20,000 people	135 litres per day	135 litres per day
Towns having population less than 20,000 people	100 litres per day	100 litres per day

Source: Study of Planning of Water Resources of Rajasthan

It is a well-established fact that water requirements for domestic uses, including those for livestock, are a function of climate and are generally high in arid tropics. The arid areas in India are also more drought-prone than semi-arid and sub-humid areas (Kumar *et al.*, 2006). Hence, the norm of having a high per capita water supply for drought-prone areas has some scientific basis. Given the fact that a large part of Rajasthan is arid and the rest semi-arid, the per capita water supply in rural areas as per the existing norms, that is, 40 lpcd and 70 lpcd, for both drought-prone areas and non-drought-prone areas is insufficient to take care of the water needs of the livestock population, which is the backbone of the rural economy of Rajasthan.

Nevertheless, the minimum quantity of water to be supplied in urban and rural areas shall be as per the new guidelines issued by the CPHEEO and NRDWP, respectively, with upward revisions, if required, as per the guidelines issued by the concerned Ministry at the Union level from time to time.

4.1.2. Quality of Water for Domestic Water Supplies

Water from the natural system may contain dissolved and suspended particulate matter, affecting its quality. Water naturally contains a wide range of living organisms, many of which are not visible to the naked eye because they are microscopic. Surface water almost always contains microbiological contaminants, some of which could be disease-causing or even deadly. This is also true of groundwater from shallow wells (generally those that are 30 m or less in depth), which can be under the influence of surface water. Water from deeper wells may also contain microbiological contaminants such as bacteria and viruses. The naturally occurring bacteria are normally not harmful, but viruses can be harmful, and aquifers and wells can become contaminated. Hence, there is a need for the treatment of raw water before it is supplied by the water supplies (both public and private).

As discussed earlier, in Rajasthan, the groundwater, which is a major source of water for domestic uses in rural and urban areas, has concentrations of salinity, nitrates, chlorides and fluorides above permissible levels in large parts, especially in the western parts of the Aravalli ranges. This water is therefore not potable unless given proper treatment for the removal of these dissolved salts. The nature and extent of treatment of the raw groundwater depend on the drinking water quality standard accepted by the state.

The nature and form of drinking water standards may vary among countries and regions. There is no single approach that is universally applicable. It is essential in the development and implementation of standards that the current and planned legislation relating to water, health and local government be taken into account and that the capacity to develop and implement regulations be assessed. Approaches that may work in one country or region will not necessarily be applicable in other countries or regions. It is essential that each country review its needs and capacities when developing a regulatory framework. However, India has its own regulatory framework, and the Bureau of Indian Standards (BIS) specification IS: 10500–1991 drinking water quality standards can be added to the existing criteria.

However, in India, so far there is no notification to make drinking water quality parameters and standards legally enforceable both in urban and rural areas. The BIS IS: 10500 drinking water quality standards are voluntary in nature. There is a need for the requirement of a separate Drinking Water Quality Assessment Authority to better focus on drinking water quality, particularly in the water scarce regions. The general criteria for assessing the quality of drinking water for organized community water supplies are given in Table 2.5. The drinking water quality standards as per BIS IS: 10500 are given in Table 2.6.

4.1.3. Frequency of Water Supplies

The international norm for water supply for domestic uses is 24 × 7 for both rural and urban areas. There is no city or town in Rajasthan that receives continuous water supply from the public system, as is the case with most Indian cities and towns. The domestic water supply provided by PHED through the schemes directly managed by it is intermittent. So is the case with the water supply schemes in rural areas, which are managed either by the PHED or by the Gram Panchayats. Intermittent water

Table 2.5: General Quality Criteria (Primary Parameters) for Raw Water for Organized Community Water Supplies (Surface and Ground Water)

	Water Quality Parameters	Range/Limiting Value of Water Quality		
		High	*Medium*	*Poor*
1	pH	6.5–8.5	6–9	6–9
2	Colour, Pt scale (Hz units)	<10	<50	<500
3	Total suspended solids (mg/l)	<1,000	<1,500	<2,000
4	Odour dilution factor	<3	<10	<20
5	Nitrate (mg/l)	<50	<50	<50
6	Sulphates (mg/l)	<150	<250	<250
7	Chloride (mg/l)	<200	<300	<400
8	Fluoride (mg/l)	<1	<1.5	<1.5
9	Surfactants (mg/l)	<0.2	<0.2	<0.2
10	Phosphates (mg/l)	<0.4	<0.7	<0.7
11	DO (% saturation)	60–110	80–120	90–140
12	Biochemical oxygen demand (mg/l)	<3	<5	<7
13	Total kjeldahl nitrogen (mg/l)	<1	<2	<3
14	Ammonia (mg/l)	<0.05	<1	<2
15	Total coliform (MPN/100 ml)	<500	<5,000	<50,000
16	Faecal coliform (MPN/100 ml)	<200	<2,000	<20,000
17	Faecal streptococci (MPN/100 ml)	200	1,000	10,000

Source: Central Pollution Control Board

supply causes problems of excessive drops in pressure in the system due to illegal tapping of water. It also causes contamination of water due to the entry of sewage through the points of leakage in the distribution pipelines. For the same amount of water to be supplied, intermittent supply means a higher carrying capacity of the pipelines carrying water and higher infrastructure costs. As the supply of water is limited consumers are forced to keep intermediate storage systems for storing the supplied water.

This is one of the ways in which consumers maintain a 24 × 7 domestic water supply at the household level in rich localities, whereas the households in poor localities have to wait for the water supply schedules to collect the required quantum of water for their daily needs, often with significant opportunity costs. Therefore, to maximize the welfare effects of improved water supply in terms of quantity and quality, it is important that water be supplied every day and be available for twenty-four hours. However, this can only be achieved progressively.

4.1.4. What Is Safe Water?

Some of the definitions for safe water as devised by various international bodies and agencies are as follows:

1. As per the WHO Guidelines for Drinking Water Quality, safe drinking water is defined as one that doesn't represent any significant risk to health over a lifetime of consumption, including different sensitivities that may occur between life stages.
2. To be safe for human consumption, drinking water must not contain disease-causing microorganisms (bacteria, viruses or parasites) or chemicals at potentially harmful levels. The physical quality ('appearance') of the water should be good, that is, it should have no suspended material such as clay or silt, and it should be clear and colourless, with no unpleasant taste or odour (Victorian Government Department of Health, 2009).
3. Safe water means water that will not harm you if you come into contact with it. The most common use of this term applies to drinking water, but it could also apply to water for swimming

or for other domestic uses. To be safe, the water must have sufficiently low concentrations of harmful contaminants. The harmful contaminants include: disease-causing microbes such as bacteria, viruses, and protozoans; cancer-causing chemicals such as many pesticides, organic solvents, petroleum products, chlorinated by-products of the disinfection process, and some metals and metalloids; nitrates and nutrients; endocrine-disrupting compounds; strong acids and strong bases; radionuclides; and any other acutely toxic substance (source: US Geological Survey).

4. Safe drinking water is water with microbial, chemical and physical characteristics that meet WHO guidelines or national standards on drinking water quality (source: WHO/UNICEF Joint Monitoring Programme).

Besides this, the WHO/UNICEF Joint Monitoring Programme (JMP) for water supply and sanitation has established a standard set of drinking-water and sanitation categories that are used for monitoring purposes, viz., improved drinking water sources and unimproved drinking water sources. An 'improved' drinking-water source is one that, by the nature of its construction and when properly used, adequately protects the source from outside contamination, particularly faecal matter. As per this classification, improved sources include: piped water into a dwelling, yard or plot; public tap or standpipe; tube wells or boreholes; protected springs and dug wells; and rainwater. Unimproved sources, which are common in Rajasthan, include: unprotected dug wells; carts with a small tank or drum; tanker trucks; and surface water in ponds, tanks, nadis and lakes (*talabs*).

4.1.5. Bacteriological Quality of Drinking Water

4.1.5.1. WATER IN THE DISTRIBUTION SYSTEM

1. Throughout any year, 95 per cent of samples should not contain any coliform organism in 100 ml of water.
2. No sample should contain E. coli in 100 ml of water

Table 2.6: Water Quality Requirements for Drinking Purposes under ISo 10500:1991

	Substance or Characteristics	Desirable Limit	Permissible Limit in the Absence of Any Alternate Source
A] Physical Requirements			
1	Colour (Hazen units, Max)	5	25
2	Odour	Unobjectionable	–
3	Taste	Agreeable	–
4	Turbidity (NTU, Max)	5	10
5	Total dissolved solids (mg/l, Max)	500	2,000
6	pH value	6.5–8.5	No relaxation
7	Total hardness (as $CaCo_3$ in mg/l, Max)	300	600
B] General Chemical Substances			
1	Copper (as Cu in mg/l, Max)	0.05	1.5
2	Iron (as Fe in mg/l, Max)	0.3	1.0
3	Manganese (as Mn in mg/l, Max)	0.1	0.3
4	Nitrate (as NO_3 in mg/l, Max)	45	No relaxation
5	Fluoride (as F in mg/l, Max)	1.0	1.5
6	Zinc (as Zn in mg/l, Max)	5	15
7	Aluminium (as Al in mg/l, Max)	0.03	0.2
8	Chlorides (as Cl in mg/l, Max)	250	1,000
9	Selenium (as Se in mg/l, Max)	0.01	No relaxation
10	Sulphate (as SO_4 in mg/l, Max)	200	400 (This may be extended up to 400, provided that Mg does not exceed 30.)
11	Alkalinity (mg/l, Max)	200	600

Table 2.6: Continued

	Substance or Characteristics	Desirable Limit	Permissible Limit in the Absence of Any Alternate Source
12	Calcium (as Ca in mg/l, Max)	75	200
13	Magnesium (as Mg in mg/l, Max)	30	100
14	Residual and Free chlorine (mg/l, Min)	0.2 (This limit is applicable only when water is chlorinated. When protection against viral infection is required, it should be a minimum of 0.5 mg/l.)	–
15	Phenolic compounds (as C_6H_5OH in mg/l, Max)	0.001	0.002
16	Mineral oil (mg/l, Max)	0.01	0.03
17	Anionic detergents (as MBAS in mg/l, Max)	0.2	1.0
18	Boron (as B in mg/l, Max)	1	5
C] Heavy Metals and Pesticides			
1	Mercury (as Hg in mg/l, Max)	0.001	No relaxation
2	Cadmium (as Cd in mg/l, Max)	0.01	No relaxation
3	Arsenic (as As in mg/l, Max)	0.01	No relaxation
4	Cyanide (as CN in mg/l, Max)	0.05	No relaxation
5	Lead (as Pb in mg/l, Max)	0.05	No relaxation
6	Chromium (as Cr^{6+} in mg/l, Max)	0.05	No relaxation

(*continued*)

Table 2.6: Continued

	Substance or Characteristics	Desirable Limit	Permissible Limit in the Absence of Any Alternate Source
7	Polynuclear Aromatic Hydrocarbons (as PAH in mg/l, Max)	NS	NS
8	Pesticides (mg/l, Max)	Absent	0.001
D] Radioactive Materials			
1	Alpha emitters (Bq/l, Max)	–	0.1
2	Beta emitters (pci/l, Max)	–	1

3. No sample should contain more than 10 coliform organisms per 100 ml of water
4. Coliform organisms should not be detectable in 100 ml of any two consecutive water samples.

4.1.5.2. UNPIPED WATER SUPPLIES

The objective should be to reduce the coliform count to less than 10 per 100 ml of water sample, but more importantly, to ensure that the faecal coliform is all together absent.

5. Water Pricing

Efficient pricing of water plays an important role in regulating the water use and ensuring equitable water distribution, efficiency of the system and its proper operation and management. Various water policies in India (of 1987, 2002 and 2012) have emphasized the importance of levying appropriate water charges for different uses of water, and the most recent NWP

(2012) has even recognized water as an economic good and recognized the need for efficient pricing of water to reflect on the scarcity value of the resource. However, the approach adopted by the state governments in fixing water prices is not uniform and characterized by subsidies, no price revision to adjust for inflation and poor revenue recovery (CWC, 2010). As a result, water in most Indian cities and towns is under-priced, with damaging long-run consequences for households who have limited and poor-quality water services and for water-supplying entities that are unable to invest and expand water coverage (Mathur and Thakur, 2003).

The mode of pricing of water – the way water supply utilities have to recover the cost of supplying water to consumers – depends on the objective, that is, whether to improve the efficiency of water use, promote equity in water allocation, improve the financial sustainability of the utility, achieve environmental sustainability, or achieve all of these.

In Rajasthan, present revenue from drinking water supply schemes is about 20 per cent of the O&M costs. Power consumption alone costs three times as much as revenue collection. The difference in unit cost of supplying water compared to user charges has led to a cycle of higher subsidies (estimated at 74 per cent) to consumers with higher consumption and likely higher wastage. The Public Health and Engineering Department (PHED) spent three times its revenue to meet the rising demand for drinking water between 1991 and 2000. The revenue-expenditure gap has risen over the years. Similar is the story in most of the other Indian states. However, of late, water supply agencies in some states have taken corrective measures. For instance, the Bangalore Water Supply and Sewage Board (BWSSB) has revised rates by almost 20 per cent every year since 1991 to enable operations on a no-loss, no-profit basis (Hooda, 2013). The Delhi Jal Board has substantially revised water charges to cover the supply cost, promote efficient water use and reduce wastage.

5.1. Criteria and Norms for Fixing the Water Tariff

The internationally accepted criterion for charging for water supplies is a volumetric tariff, when water is scarce and its production and supply costs are high (ADB, 2007; World Bank, 2007). The underlying principle is that there is a positive marginal cost of increasing production and supply, which comes from the cost of sourcing and supplying water[1], the cost of resource depletion and the cost of environmental degradation (World Bank, 2007). This has inherent concerns about equity. There are other reasons why volumetric pricing is desirable over flat rate pricing. Flat rate pricing creates a demand to expand water delivery beyond the efficient point. If the costs of installing and running the water system are not primarily financed by water revenues but by general taxes, underpricing water can be very expensive to society as it would mean massive, uneconomic expansion of the water delivery system (Noll *et al.*, 2000). Pro rata tariffs send the right price signals for consumers to realize the scarcity value of the resource, thereby reducing production externalities. Also, for financial prudence, it is important that the utility recover the charges from the consumers on the basis of the volume of water used and the average unit cost.

When water is scarce, the marginal cost of production and supply of water can also increase significantly at times when the cost of augmenting the supplies goes up exponentially with every unit increase in the volume of fresh water, owing to the need for additional infrastructure for water transfers or the addition of a new water treatment system (say, a desalination plant). As pointed out earlier, a pro rata tariff reduces resource depletion and environmental degradation costs through efficient water use (Kumar, 2014) without compromising on the developmental outcomes of improved water supply, thereby reducing the average unit cost. For a greater reduction in water demand and to bring down average unit cost, an incremental block tariff (IBT) can often be used by the utility with higher unit rates for a higher volume of consumption (Olmstead *et al.*, 2007).

Whereas, when the resource (i.e. water) is abundant, the resource (depletion) cost and environmental degradation cost would be ideally low. Under such circumstances, the (fixed) cost of the infrastructure for treating and supplying water would constitute a major proportion of the total cost of water supplies. In that case, normally, lowering unit rates or

1 The cost of the water supply infrastructure, and operation and maintenance costs.

a flat rate system of pricing water is followed to encourage people to use more water and, in the process, increase the capacity utilization of the existing infrastructure, thereby being able to reduce the unit cost of production and supply of water.

Other criteria for fixing the water tariff in domestic water supplies include affordability and cost recovery. As drinking water supply is a social good and helps meet developmental objectives of reducing waterborne diseases, malnutrition and infant mortality, affordability is an important concern. However, any attempt to increase affordability by lowering the water tariff would affect the financial workings of the utility. Therefore, targeted subsidies are often resorted to for improving affordability amongst poor households. In certain other situations, a low unit rate is charged for the consumption of a certain quantum of water per family in a month if it is within the basic minimum volume of water required for survival to avoid market failure (Noll *et al.*, 2000), as it produces the highest social returns. Such an approach reduces the transaction costs involved in introducing targeted subsidies, but in the process, the benefits of the subsidy would also be passed on to rich households.

The issue of cost recovery can be addressed by adopting either of the two following strategies. The first one is by charging higher rates than the average cost of production and supply of water from consumer segments that use water for commercial gains like hotels, restaurants, parks and other commercial establishments, and charging very low unit rates for the domestic sector, where large numbers of poor households are potential consumers. This is a strategy followed in many Indian cities (ADB, 2007). The reason is that affordability may not be a concern for the former, and returns from such uses are not likely to be high for society at large. This is 'cross subsidization', in the sense that at the aggregate level, the utility would be able to recover the cost of water supply, while the high priority uses enjoy access to water at affordable prices. In water-rich/abundant areas, since flat rates are the ideal mode of charging for water, these consumer segments can be charged a high 'slab rate', high connection charges or both.

The second method for achieving cost recovery is by adopting incremental block rates for all consumers (IBT) with high unit rates for higher levels of per capita/ per household consumption in excess of the basic minimum[2], which do not necessarily produce high social returns. However, the second method will be socially viable and politically acceptable only in water-scarcity situations. In urban areas, this approach would face a new challenge in implementation, as bulk water supply to residential complexes is common, where water is delivered to a common point. In the case of bulk water supply, strategy has to be devised in such a way that the basic minimum will have to be kept at very high levels, depending on the number of households within the residential complex, and the additional slabs are also worked out keeping in view these households.

The most common mode of pricing of water by the PHED in rural and urban areas of Rajasthan is the water charge based on number of household connections in rural areas and the size of the household/dwelling connection in urban areas. There is a PHED notification on water tariff for consumers served in urban, rural and special project areas, dated 29 May 1998, and a subsequent amendment to this notification, which stipulates the penalty for illegal water tapping and the levy of an infrastructural development (for water) surcharge, dated 17 September 2007. Metering and volumetric pricing are followed only in a few cities in Rajasthan, and those too only cover the consumer base partly. Further, currently, there is no system of metering water supply to households in rural areas.

The criteria and norms proposed for pricing of water supply are based on the established theory that the price elasticity of water demand is negative (Olmstead *et al.*, 2007; Renwick *et al.*, 1998), the price elasticity of water demand increases under increasing block rates (Olmstead *et al.*, 2007) as compared to uniform volumetric rates, and for higher income groups, the price elasticity of water demand would be lower as compared to lower income groups (Foundation for Water Research, 2004). In other words, the demand reduction through a given hike in the water price would be less for higher income groups as compared to lower income groups. Also, a study by Renwick *et al.* (1998) indicated that policy instruments such as public information campaigns, retrofit subsidies, water use restrictions and rationing are capable of reducing residential water usage. However, mandatory policies, such as use restrictions, reduced demand more than voluntary

2 Essentially for gardening, car washing, swimming pools, etc.

measures, such as public information campaigns, as indicated by higher values of estimated coefficients of demand elasticity for these measures.

6. Private Sector Involvement in Water Supply

In an attempt to shift towards water privatization during the economic liberalization period of the 1990s, a set of measures favouring total reduction in subsidies, full cost recovery, and privatization of water supply services has been promoted around the world. The Dublin Conference on Water and Environment in 1992 unfolded a new approach to water resources management that encouraged the privatization of the water sector in developing countries. It declared that water has an economic value and must be treated as an economic good. It also stated that access to water and sanitation are fundamental human rights. The fourth principle of the Dublin Declaration states 'Water has an economic value in all its competing uses and should be recognized as an economic asset'. Following these principles, it is crucial to recognize the basic right of all human beings to have access to drinking water and sanitation at an affordable price.

Past failure to recognize the economic value of water had led to wastage and uses that were harmful to the environment. To manage water as an economic asset is the key to efficient and equitable use and promoting conservation and protection of water resources (Petrella, 2001). Following the structural adjustment programmes in India since 1991, the Eighth Plan (1992–1997) made a significant departure from the past by giving a thrust towards the privatization of the water sector. The main thrust and strategies were, among others, to: manage water as a commodity in exactly the same way as any other resource; encourage the private sector for construction and maintenance of drinking water projects and mobilize them to the maximum extent feasible; and ensure that, in urban areas, municipalities are free to levy and raise appropriate user charges for drinking water and sanitation facilities in order to strengthen their financial position. The Tenth Plan (2002–2007) also further stressed the philosophy of liberalization in the water sector: water needs to be managed as an economic asset rather than a free commodity, like any other resource, and the supply of water to consumers should normally be based on the principle of effective demand, which broadly corresponds to the standard of service that the users as a community are willing to maintain, operate and finance. The National Water Policy (2002) of the Government of India has reflected the changing paradigm in the water sector and included encouragement of 'private sector participation in the planning, development and management of water resources projects'.

In the last couple of decades, there has been enormous debate in India on the role of the private sector in providing water supply. Advocates of public-private partnerships (PPP) argue that increasing private sector involvement would solve many problems plaguing water and sanitation utilities, including their failure to provide services to the urban poor. On the other hand, critics argue that increasing private sector participation was part of the problem, another step in the dismantling of the policies and institutions needed to achieve universal coverage of adequate water and sanitation. In the 1990s, proponents of private sector involvement in water supply launched a sustained critique of public utilities and their failures and promoted a (regulated) private alternative. Particularly in low-income settings, it was argued, public utilities are inclined to be inefficient, overstaffed, susceptible to corruption, open to manipulation by politicians pursuing short-term political ends, and unresponsive to consumer demands. Low water tariffs, far from ensuring that low-income households can afford piped water, turn water distribution into patronage and contribute to utilities' financial difficulties, often inhibiting investment and preventing water and sanitation networks from being extended to low-income settlements (even when residents are willing to pay). Private utilities, proponents believe, would be cost-conscious, apolitical and responsive to demand.

Despite being contentious, it is generally believed that the presence of independent regulators working in the public interest, offering competition for concessions or other contracts, will reduce the monopoly power of private utilities to exploit consumers, while tariff reform would ensure cost recovery, thereby achieving what public utilities failed to achieve.

7. Regulations on Private Water Supply Service Providers

Private operators usually use less technology and charge rates much higher than those charged by the water utility. They survive because they are flexible in their approach and responsive to consumer demands. They provide assured services and, in some cases, are even willing to change payment patterns to suit their customers. In India, water supply is a mandatory function of urban local bodies. However, the agencies responsible for water supply are unable to serve the entire population within their jurisdiction due to numerous problems: lack of funds, inadequate infrastructure, population growth, inefficiency in provision, high levels of unaccounted water pollution and source depletion. The net result is that citizens receive inadequate water from public agencies and have to find their own solutions to tide over the crisis. This situation creates a conducive environment for small-scale private water vendors to operate. These private operators bridge the gap between demand and supply and provide water to unserved and water-scarce areas. They operate informally, and their main business is supplying water. They invest private funds in the business, and given the water supply situation in many Indian cities, they recoup their investment within a few years.

The number of private water vendors has increased in rural and urban areas of Rajasthan in the recent past. For instance, in Jaipur city alone, about 1,640 trips per day are made by tankers to meet the water supply-demand gap (source: PHED). There are mainly four types of private tanker operators in urban areas of Rajasthan: private tube well owners and tanker contractors; private tanker owner-operators; PHED contract tankers; and disguised PHED tankers. The price at which water is sold by the tanker operators varies from INR 63 to INR 65 per kiloliter (Birkenholtz, 2010). There is an urgent need to provide the necessary rules for regulating their functioning so as to achieve better developmental outcomes in terms of consumer satisfaction and protection of their rights. First of all, certain rules have to be framed on who can become water vendors, the 'quality of water' that needs to be maintained for various purposes (drinking and non-drinking), and consumer responsibilities. Also, the functioning of private water vendors will have to be monitored periodically, and action will have to be taken against the vendors who violate the rules. The Drinking Water Supply Regulations Act should clearly spell out who can monitor, the penalties for non-compliance, etc.

One such initiative is taken by the Food Safety Commissioner of Kerala, which has come out with a special set of guidelines to ensure the supply of drinking water through tankers. The norms make it mandatory for tankers to possess a Food Business Operator (FBO) license under the Food Safety and Standard (Licensing and Registration of Food Business) Regulations, 2011. As per the guidelines, the FBO license number should be displayed on the tankers. Moreover, establishments using multiple vehicles to supply water are required to obtain a special license that contains the registration numbers of all vehicles.

All private drinking water sources must have an FBO license. The vehicles used for supplying drinking water must carry signage in English and Malayalam. For those vehicles transporting water for other purposes, the signage must state 'Not for Drinking Purpose' on the tanks. Legal action would be initiated against those transporting water without the proper signage. The quality of the water should be ascertained every six months by testing at government laboratories or those accredited by the National Accreditation Board for Testing and Calibration Laboratories. The norms also specify that the interiors of the water tanks must be coated with certified materials in order to prevent contamination of the water.

The vehicles transporting drinking water for supply are required to have a Food Safety and Standards Act (FSSA) license, a laboratory report stating the water quality, and a certificate regarding the coating material used in the tank. Failure to produce the required documents is a punishable offence under the FSSA, 2011.

The norms also direct the public to purchase water only from those distributors with valid FSSA licenses. Hotels, restaurants, apartments, hospitals and other establishments that purchase water have to maintain a register that contains the details of the supplier. It should also contain the water source, copies of the lab reports on the quality of the water, license details of the supplier, and a copy of the agreement for the water supply and the quantity of water. Non-compliance with the regulations would invite legal action under the FSSA.

8. Rules and Regulations Pertaining to 'Illegal Water Tapping' and Source Protection

Illegal water use is a serious problem for many small urban utilities. It takes many forms, including unauthorized connections to the water network and meter tampering. Illegal water use reduction cannot be handled in an ad-hoc way but should form an important component of a utility's programme to reduce UFW (unaccounted for water) and maximize revenues and operating efficiencies. A clear and practical Illegal use reduction policy can help in this regard. For instance, in Uganda, any household or person tapping water in an illegal manner is fined USD 200 (UN-Habitat, 2012). Similarly, initiates are clearly lacking in most of the Indian states.

There are huge commercial losses to urban water supply utilities in Rajasthan on account of distribution losses, illegal water tapping, non-functional meters and un-metered water supply connections. The unaccounted-for water (UFW) varies from 62 per cent in Kota to 24 per cent in Bhilwara city (Mittal, 2004). A study assessing UFW in Jaipur city found that almost 50 per cent of the consumer meters were non-functional (Grieve and Johnson-Jones, 2000). Mittal (2004) highlighted that illegal water connections are a major cause of water theft in the urban areas of the state and are thus mainly responsible for the huge gap between the volume of water supplied and the metered consumption.

In addition to illegal water use, activities that threaten the safety and sustainability of drinking water sources through contamination and illegal water tapping are also rampant in Rajasthan. The sources that are subject to such a threat include surface water bodies such as lakes, tanks, reservoirs and freshwater aquifers. In rural areas, it is a common phenomenon that wells are drilled by farmers in the vicinity of groundwater-based drinking water sources, created by the water supply agency after thorough scientific investigations to assess resource availability. These agro-wells interfere with the drinking water wells, resulting in their yield reduction. The Act should be comprehensive enough to have stringent provisions to deal with all such situations, wherein drinking water sources are contaminated, water is tapped illegally from such sources for agricultural and industrial uses, or water is tapped illegally from the public water supply system using un-authorized connections for domestic use.

9. Technology Choices for Water Supply

In lieu of the fact that Rajasthan faces acute water shortages for domestic water supply, induced by the availability and quality of water in the natural system, especially in aquifers, there is a need for employing advanced technologies like desalination systems for creating new water, generally known as 'technology water'. However, currently, the technologies being attempted in the drinking water sector are conventional, such as tube wells, open wells, hand pumps, reservoirs, lakes and 'diggies'.

Poor selection of technology can threaten the sustainability of water supply services from physical and socio-economic points of view (UNICEF/IRAP, 2012). The technology choice mostly depends on the basic strategy adopted by planners and on general trends in the water and sanitation sector. However, in a water-scarce region like Rajasthan, the technology choice should be driven by hydrologic (water availability, water quality, and dependability) and economic considerations to ensure sustainability of water supply services in lieu of growing scarcity, groundwater depletion, environmental water stress in rivers and the increasing cost of production and supply of water.

As noted by Reddy et al. (2010), the failure of village water supply schemes based on bore wells is very high in India, especially those in hard rock areas. The high 'slippage' witnessed by many states that depend on bore wells for village water supply (Chaudhary et al., 2020; Reddy et al., 2010) made a compelling case for looking at the life cycle cost of schemes that otherwise are not capital-intensive (Reddy *et al.*, 2012). Concerns about the physical sustainability of the source should force the water supply planners to look for technical options that might be capital-intensive, but sustainable in the long term, thereby offering cost effectiveness. Such concerns are relevant for Rajasthan, given the fact that nearly half of the geographical area of the state is underlain by hard rock formations, and in a major part of the remaining area, chemical contamination of groundwater (with salinity, fluoride and nitrates) is extensive, and yet nearly 61 per cent of the villages are covered by groundwater-based schemes (with bore wells, tube wells and hand pumps) (see Table 2.2 of this chapter).

There are several factors that influence the selection of community-based water supply technology. Some of the factors of general relevance are presented in Table 2.7.

The above listed criteria along with the considerations for (physical and economic) sustainability of services should form the basis for deciding the criteria for selection of water supply technologies in the State of Rajasthan.

10. Rainwater Harvesting

Rooftop rainwater harvesting is being promoted in many areas in India to improve domestic water security (Arabindoo, 2011; Kumar, 2004, 2009). However, the most important factors influencing the performance of rooftop rainwater harvesting systems are the quantum of annual rainfall, the number of rainy days, the per capita roof area, the topography and the population density of the area under consideration. Kumar (2009) found that the physical performance of the system is better when the annual rainfall is high (above 2,000 mm) and the roof area is large (especially in the case of bungalows). The economic viability improves when the rainfall is distributed over long time periods – a characteristic of high rainfall regions – in which case the effective storage could be high with low storage volumes, reducing the cost of the roof water collection tank. Economic viability will also be high in both hilly regions and regions with very sparse populations, where conventional water supply systems

Table 2.7: Factors of General Relevance Influencing Selection of Community Water Supply Technology

Factors	Criteria
Technical Factors	1.1. demand (present and future consumption patterns) versus supply 1.2. capital costs 1.3. extension capacity 1.4. compatibility with norms and legal frameworks 1.5. compatibility with existing water supply systems 1.6. comparative advantages 1.7. technical skills needed within, or outside, the community
Environmental Factors	2.1. availability, accessibility and reliability of water sources (springs, ground water, rainwater, surface water, streams, lakes and ponds) 2.2. seasonal variations 2.3. water quality and treatment 2.4. water source protection 2.5. risk of a negative environmental impact
Institutional Factors	3.1. legal framework 3.2. regulatory framework 3.3. national strategy 3.4. existing institutional set-up 3.5. support from government, NGOs, external support agencies (ESAs) 3.6. transfer of know-how
Community and Managerial Factors	4.1. local economy 4.2. living patterns and population growth 4.3. living standards and gender balance 4.4. household income and seasonal variations 4.5. users' preferences 4.6. historical experience in collaborating with different partners 4.7. village organization and social cohesion
Financial Factors	5.1. capital costs 5.2. budget allocations and subsidy policy 5.3. financial participation of users 5.4. local economy

Source: Adopted from Brikké and Bredero (2003)

become prohibitively expensive in terms of the cost of the infrastructure per capita. In urban areas with residential complexes, space for building the storage system can also be a constraint.

Also, if the households, which are connected to the public water supply systems, do not pay for the water on a volumetric basis, they won't have much incentive to reduce their dependence on the same even if they build rooftop rainwater harvesting systems (RRWHS) (Kumar, 2004). Therefore, any financial incentive provided to the households for building RRWHS should be subject to the condition that these households go for a metered tariff for public water supply instead of a flat tariff. These factors need to be taken into account while framing the rules for rainwater harvesting in Rajasthan so that they produce positive equity outcomes along with improved service delivery.

11. Wastewater Treatment and Reuse

Technology choice for treatment of domestic sewage eventually depends upon wastewater characteristics and on the treatment objectives as translated into desired effluent quality. Effluent quality control is typically aimed at public health protection, preservation of the oxygen content in the water, prevention of eutrophication, prevention of sedimentation, preventing toxic compounds from entering the water and food chains, and promotion of water reuse (Helmer and Hespanhol, 1997).

Options available for cost-effective and environmentally compatible sewage treatment include land treatment, waste stabilization ponds, constructed wetlands, duck-weed ponds, aerated lagoons, rotating biological contractors, up-flow anaerobic sludge blanket systems, root zone treatment (CPCB, 2008; Kumar, 2014) and geo-environmental treatments such as soil aquifer treatment (SAT) (Kumar, 2014).

CPCB (2005) notes that for large cities where land availability is limited, treatment schemes based on primary sedimentation followed by an activated sludge process with an anaerobic sludge digester and sludge drying beds for anaerobic sludge are quite suitable. For small cities/towns where land can be made available, treatment schemes based on grit removal followed by an up-flow anaerobic sludge blanket reactor and further followed by a polishing pond are suitable technologies. Whereas, for small towns where sufficient land is easily available, a treatment scheme based on a series of waste stabilization ponds is the most economical and appropriate technology (CPCB, 2005; Kumar, 2014). Multiple stage ponds (at least three), first pond being an anaerobic one, are the most widely used and suitable configuration (CPCB, 2008). Climate is an important factor determining the performance of most wastewater treatment systems (aerobic and anaerobic), and therefore this factor should be taken into account while choosing the type of treatment technology. So are the availability of human resources to run the treatment plants and the potential for reuse of the treated water, as they would increase viability (Kumar, 2014).

In the context of Rajasthan cities/towns, the treatment systems for cities and towns will have to be chosen by keeping in view the following: the nature of the city (like corporation, municipality, municipal towns, etc.), which would determine the institutional viability; the degree of urbanization, which would determine the price of the land; climatic conditions; and the geo-hydrological environment.

The potential for reuse of treated wastewater already exists in Rajasthan because, in most of its regions, irrigation water scarcity exists and the treated wastewater can be diverted for crop production. Therefore, the level of treatment of wastewater can be such that it conforms to effluent quality guidelines for human health and crop protection stipulated by the WHO (1989). Provisions need to be made in the Act to improve the technical performance and socio-economic viability of wastewater treatment technologies.

There may be situations where there is no demand for treated wastewater for use in irrigation. In such a situation, the treated wastewater will have to be discharged into natural watercourses. In such cases, the wastewater should be treated to such standards for bathing and propagation of wildlife and fish populations, prior to discharge, as stipulated by the WHO (1989).

Since the state is 'physically water scarce', with the average per capita renewable water resource availability just above the threshold of 500 m^3 per annum for becoming 'absolutely water scarce', in the case of freshwater being used for industrial production, it is necessary that the effluent is treated to tertiary levels so that it doesn't put additional stress on the limited fresh water resources in the state.

12. Water Supply Surveillance

For the improvement of quality of life and enhancement of productive efficiency among the inhabitants, there is a need for very good water supply and sanitation provision in both rural and urban areas. Water is tapped and supplied for domestic and industrial uses from rivers, streams, wells and lakes. However, the reliability of the water supply mainly depends on the surveillance and effective management of the available water resources. Thus, water supply surveillance is designed to protect public health through the identification of inadequacies and by promoting the timely implementation of action to control risks (Howard, 2001). Surveillance is an exploratory action that is intended to identify and evaluate factors associated with drinking water that could pose a serious health risk.

The objective of a surveillance programme is to contribute towards the protection of public health by improving the quality, quantity, coverage, cost and continuity of water supplies (WHO, 1997, NDMA, 2010), and not merely to collect and collate information (NDMA, 2010).

Surveillance requires a systematic programme of surveys that combine analysis, sanitary inspection, and institutional and community aspects. In most countries, the agency responsible for the surveillance of drinking-water supplies (urban and rural) is the ministry of health or its regional or departmental offices. In some countries, there is an environmental protection agency, and in others, environmental health departments of local governments may have some responsibility. The surveillance agency should preferably be an established national institution designated by appropriate legislation and should be able to operate at the central, provincial and local levels (WHO, 1997).

Further, water quality surveillance requires an appropriate institutional framework and adequate resources (financial, infrastructure and human) to function effectively. Surveillance activities need to be adapted to local conditions and the availability of local financial resources, personnel, infrastructure and political commitment.

Surveillance must therefore take into account the fact that a wide range of factors influence the health outcomes of water supply and that a broad-based system of data collection and interpretation is needed. Surveillance programmes are also designed to provide independent verification of the degree to which the water sector is meeting the basic right of people to a safe and adequate water supply. Access to a safe and adequate water supply may be restricted due to prohibitive charges, daily or seasonal fluctuations in availability or failure to extend supplies to low-income or informal settlements. The cost of water purchased from neighbours or from public facilities may represent a significant proportion of family income and limit the amount of water that may be collected each day. The continuity of supply may be poor due to weak O&M of the supply, thus forcing families to use reduced volumes of water and store water. Elsewhere, seasonal, geographical and hydrological factors may deny individual households or entire communities a continuous, reliable supply of drinking water. During dry seasons, spring sources may dwindle, reservoirs may empty, and excessive demands by one group may compromise the needs of their neighbours (Howard, 2001).

Surveillance is complementary to the quality control function of the drinking-water supply agency. Surveillance does not remove or replace the responsibility of the drinking-water supplier to ensure that a water supply is of acceptable quality and meets pre-determined health-based and other performance targets (WHO, 2008).

A drinking-water supply surveillance programme should normally include processes for the approval of Water Safety Plans (WSPs). This approval will normally involve a review of the system assessment; the identification of appropriate control measures and supporting programmes; and operational monitoring and management plans. The surveillance agency may also support or undertake the development of WSPs for community-managed drinking-water supplies and household water storage. Such plans may be generic for particular technologies rather than specific for individual systems (WHO, 2008).

13. Capacity Building in the Water Supply Sector

Capacity building is the process of gaining technical, managerial and institutional knowledge and insight in relation to the socio-economic structure, cultural standards and values of the society concerned. It also involves skill development. The fundamental goal of capacity building is to enhance the ability to evaluate and address the crucial questions related to policy choices and modes of implementation among development options, based on an understanding of environmental potentials and limits and on needs as perceived by the people of the country concerned. The Delft Declaration at the UNDP symposium in 1991 describes capacity building as a global concept and a strategic element in the sustainable development of water resources and identifies the three basic elements of capacity building as:

1. Creating an enabling environment with appropriate policy and legal frameworks;
2. Institutional development, including community participation; and
3. Human resources development (HRD) and strengthening of managerial systems.

The UNDP's capacity building programme for sustainable water sector development stresses 'vertical' capacity building within an individual water sector and 'horizontally' between sectors. It recognizes that capacity building is a long-term and continuing process involving all stake-holders.

The past experiences in the water supply sector have shown that, institutional weaknesses and malfunctions are the major causes of ineffective and unsustainable water services. The institutional weakness often comes from poor institutional design. For instance, the State Pollution Control Boards, including that of Rajasthan, are responsible for monitoring pollution as well as enforcing the norms on effluent disposal in water bodies. This unique situation leaves least incentive among them to monitor water quality periodically and publish the same. Similarly, the line agencies such as water supply department, M & M irrigation dept. MI dept. and groundwater dept. follow a 'sectoral and segmented approach', developing their sources independently, without considering the hydrological integrity of the water resource systems. In the process, acts of one line of agency adversely affect the performance of others.

For instance, in Rajasthan, the large scale building of water harvesting structures and watershed development activities being undertaken by the State Watershed Department and the Minor Irrigation wing of the Water Resources Department (erstwhile the Irrigation Department) are found to be reducing the flow into reservoirs and lakes (Ray and Bijarnia, 2006), which are used as sources of water supply by the state PHED. The most recent case is of the Ramgarh Lake, which used to be the major source of drinking water for Jaipur city but whose upper catchment witnesses large scale encroachments by farmers for cultivation and the building of numerous Anicuts (Singh, 2012). At present, there are no provisions under the existing laws by which such activities can be stopped. There is no coordination amongst the concerned departments at the time of the planning of these schemes, and more importantly, this activity is unregulated at the basin level. Therefore, there is a need to manage overall water resources coherently and to facilitate the allocation of water among all users suggests integrated planning (Kumar, 2006; 2014).

13.1. Need for Capacity Building in the Water Supply Sector

Water resources management and water allocation are the most important challenges that are faced today in water-scarce regions for improving water-related services. There are many failures in water resources management due to a lack of trained staff and proper institutional framework (Kumar *et al.*, 2012). Ismail Serageldin, the then Vice President of the World Bank, in his address to the VIII World Congress on Water Resources of the International Water Resources Association (IWRA), held in Cairo, Egypt (1994), opined: 'How well we manage water will determine our ability to pass on our natural heritage to the future generation, maintain human health and feed a growing population'. He also highlighted dearly that 'current water policies and management practices are not sustainable from any perspective: social, economic or environmental'.

Capacity building depends on two interrelated concepts: (1) strengthening institutions at all levels to deal more effectively and efficiently with all aspects of sustainable water resources development, including the creation of a favourable policy environment, water resources assessment (both quantity and quality), planning, management, and programme and project formulation, implementation and evaluation; and (2) the development of human resources at all levels, including education, training, and the creation of working conditions conducive to job satisfaction and performance (Alaerts *et al.*, 1991).

Capacity building seeks to integrate the management of the several subsectors in the water resources sector, particularly water supply, irrigation and environment – which often compete with each other (Alaerts *et al.*, 1991). This is particularly important for Rajasthan given the growing competition between the first two, which often lead to conflicts. The demands for water for urban growth as well as irrigation will require substantial investments in projects. But the technical conditions under which systems must be planned in developing countries are often more difficult than those in industrialized countries. Highly variable geographic and climatic conditions limit the use of standardized design for the project, and most of the projects are site-specific (Alaerts *et al.*, 1991).

13.2. Institutional Change

The most essential part of capacity building is institutional change. There is no universally accepted model for institutional change. Institutions in the water sector include national and administrative agencies such as ministries, local administrative agencies such as regional water authorities, municipal departments of public works and irrigation districts, regulatory agencies such as water pollution control and environmental protection agencies, and utility organizations. Community consumer groups are considered separately (Alaerts *et al.*, 1991).

It is well recognized that the problems associated with the water supply are more institutional than technical in nature. If the institutional arrangements can be improved, there is definitely a possibility for better improvement in the water supply. Therefore, it is essential to have the right institutions for the delivery of water supply services. However, different contexts have variable institutional arrangements, peculiarities and capacity-building needs. Thus, it is important to conduct a holistic assessment of the existing capacity to systematically design and develop the framework required for the institutional change process (Bassi and Kumar, 2012).

Institutional change is a prerequisite for achieving higher efficiency, transparency, accountability and sustainability in the water supply sector. However, in the past, capacity-building efforts were typically implemented as training and education programmes based on the idea that equipping individuals with new knowledge, skills and professional competencies would enable them to successfully operationalize sustainable practices. Many studies in the recent past, both in developed and developing countries, on the restructuring of water institutions suggest an emerging global consensus on some key principles of institution reforms.

The key principles of institutional reforms include: (1) decentralization of responsibility to the lowest appropriate levels of government to respond to local conditions and needs; (2) building of autonomous utilities with commercial orientation and financial viability for service delivery in urban areas; (3) private sector participation in the delivery of water-related services, including that with small scale providers; (4) and community participation in water management. The process of formulating the drinking water services Act for the state should explore the various provisions required for all four. Among these, decentralization of responsibilities to the lowest levels of government has already happened, with the PRIs entrusted with the job of managing water supply at the Panchayat level. The WSSO is already in place to promote community participation in drinking water management, though it needs strengthening.

The new Act will have to focus on reorienting the existing water supply institutions to improve their physical, financial, institutional and environmental performance, and setting guidelines for measuring the same. Some of the steps in institutional reforms are: (1) introduction of volumetric water pricing for water supply to consumers by the public water utility (here PHED); (2) legal powers to the SPCB to introduce penalties for non-compliance of norms on effluent quality, and recycling and reuse of treated wastewater; (3) legal powers to the SPCB to offer tax incentives to industries and the domestic sector for wastewater treatment and recycling; and (4) introduction of a water tax for the water diverted from the natural system by different users and user agencies.

These reforms would create incentives for efficient water use and reduce pollution by water users and polluters, respectively. With the introduction of volumetric water pricing, the induced marginal cost of using water from the public system would also encourage consumers to invest in technologies such as rooftop rainwater harvesting. Another possible reform measure is the separation of water-related services from water resources management functions through the creation of new organizations for water resources management. This can create an incentive among the various line agencies in water supply (irrigation, domestic water supply and industrial water supply) to engage in efficient water user practices and reduce system leakages by increasing the opportunity cost of diverting and using water from the natural system. All these measures would improve the finances of the water supply agencies and the newly created agency for water resources management.

13.3. Human Resource Development

Capacity building depends on adequate institutions, and institutions depend upon human resources (Alaerts *et al.*, 1991). With institutional change, it is also equally important to provide the necessary technical assistance to develop the local capacity of a municipality/water supplying utility for a variety of administrative, financial and managerial activities. Human resource development (HRD) is an important part of the capacity building process, which will help equip the water utility staff with the understanding, skills and access to information, as well as the knowledge and training that enables them to perform water management functions effectively. The HRD process can be undertaken by both professionally sound recruitment and providing training to existing staff to make them more competent in handling their functions. The HRD interventions should respond to local demand, which, in turn, will be developed by the opportunities created by the systemic changes and by the greater awareness generated by the dissemination of best practices.

Given the emerging challenges of enhancing the capacity of wastewater treatment systems in cities, leakage detection and reduction in water supply systems, theft prevention, fixing water tariffs, metering and recovery of water charges, and building technically sophisticated water supply systems such as desalination systems, the technical, managerial and institutional capacities of the existing line departments in the state of Rajasthan (i.e. PHED) will require strengthening. Similarly, the capabilities of WSSO will have to be strengthened to promote local institutions for the management of drinking water systems at the local level, especially in rural areas.

14. Summary and Conclusions

In this chapter, we have discussed the scenario of the drinking water sector of Rajasthan with particular reference to the water resource endowment in terms of quantity and quality, water supply systems that tap these resources to supply for domestic uses, the community's access to water supplies, the norms and criteria followed by the water administration of the state pertaining to quantity, quality and frequency of water supply for domestic purposes and pricing of water, the criteria followed for choosing technologies for water supply, the policies pertaining to rainwater harvesting for drinking water security, policies and practices for wastewater treatment and reuse, the rules and regulations pertaining to illegal tapping of water, the existing state regulations on private sector involvement in water supply sector, the status with regard to water supply surveillance, and capacity building of drinking water sector. These analyses provide the basis for deciding on the kind of provisions the drinking water services Act should have for the state.

Our analysis shows that within the water supply sector, there are some key areas where a legal/regulatory vacuum exists and where the establishment of a strong legal regime is required for improving the sector's performance. They are: selection of water supply technologies; water pricing; treatment and reuse of wastewater, especially in urban contexts; private sector involvement in water supply services; functioning of private water supply service providers; illegal tapping of water or water theft, which is rampant in urban areas; water supply surveillance; and long-term capacity building of the agencies involved in the water supply services.

CHAPTER 3

A Review of Water Services Acts in Indian States and Other Countries

1. Introduction

Since water is an essential element for life on earth, its acquisition, supply, usage and pollution have been governed by various laws since ancient times (Cullet and Gupta, 2009). The government enacts several laws to suit its needs, not by a single department but by several operational departments. This is necessary, in one way, to avoid rigidity, but the multiple enactments sometimes lack cohesion and integration. An attempt is made here to abstract several laws related to drinking water enacted by various state governments in India while also providing the scenario in select developed countries.

Water is a state subject under the Indian Constitution, though certain aspects are placed on the concurrent list (Cullet, 2016; Upadhyay, 2011). The Human Development Report (UNDP, 2006) emphasizes that water is a basic human right. Historically, the core concern of water law has been, rightly, the development principles concerning access to and control over water. Very little attention has been paid to the question of quantity and quality of supplies and the resulting equity impacts. However, of late, research studies have increasingly focused on water quality, hygiene and clean water.

The Millennium Development Goals (MDGs) enjoin signatory nations to extend access to improved sanitation to at least half of the urban population by 2015. UN water reports, since 2003, have provided updates on the status of sanitation. These reports provide a comprehensive picture as well as an in-depth analysis of specific issues or geographical areas. The fourth report (WWAP, 2012) focuses on key challenges, recent developments, emerging trends and the external drivers and pressures they place on water systems. Examples cited range from adaptive design criteria for infrastructure or demand management to institutional capacity development and policies for different developmental sectors.

The WWAP (2012) highlights that one of the limitations of the MDGs is that they neglect the centrality of water to the other goals. While the delivery of drinking water and sanitation services should remain a focus of human health and development, the importance of water resources and water governance in achieving the goals collectively needs explicit recognition. It is critical that drinking water and sanitation issues do not divert attention from the need for efforts to enhance institutional arrangements for water, with water allocation frameworks that prioritize water for these and other basic human needs above all other uses and which encourage efficiency and productivity in resource use and management.

The International Environmental Law Institute (IELI) has developed draft guidelines for the preparation of legislation pertaining to drinking water regulations for the Department of Drinking Water Supply Rajiv Gandhi National Drinking Water Mission of the Ministry of Rural Development, Government of India (IELRC, 2007). There is also a drinking water quality monitoring protocol developed by the IELRC for the Ministry of Rural Development, Government of India. There are also respective ministerial guidelines for urban water sector reforms and private sector participation. There are also guidelines for the implementation of schemes and projects on 'sustainability' under the Accelerated Rural Water Supply Programme.

Transposing the Acts and laws made elsewhere in the world, or for that matter strictly following the guidelines developed for the country as a whole, will not help either in enforcement or in bringing out the desired outcomes, without understanding the climatic, environmental, socio-economic, cultural, and political context. All these factors influence the appropriateness of certain rules and laws in a given context, be it an Act governing water supply norms, norms related to water quality, the type of technology for water supply, norms for fixing water tariffs, penalties for non-payment of water charges and illegal connections and theft, private sector involvement in water supply, or the institutional arrangement for managing water supply. It would require a lot of grounded analysis.

First of all, there is a need to clearly define 'safe water' and 'risk-free water' for consumer awareness. There is also a need to understand the critical factors that influence the per capita water supply requirements of populations, the frequency of water supply and water quality standards, including physical, chemical and bacteriological quality, in both rural and urban areas. It is also important to consider how these factors vary across different regions of Rajasthan. Accordingly, the norms related to water supply, frequency of water supply, quality of supplied water, etc., can be decided. This can help avoid developing unrealistic norms that significantly cost the state exchequer without producing any substantial public health benefits.

Similarly, there is a need to understand the factors that influence the cost of production and supply of water and the factors that influence the willingness to pay for it – two important factors that form the basis for setting the water tariff, which varies across regions. Such an understanding would help develop criteria and design the norms for setting a 'water tariff' and subsidies that integrate the concerns of cost recovery, affordability, and ease in collecting water charges. The proposed Act can take cognizance of these.

Furthermore, it is also important to understand the key (physical, socio-economic and institutional) factors that influence the performance of different types of water supply systems and the manner in which they influence them before framing rules on the selection of water supply technology for an area. Only such understanding can ensure the sustainability of the chosen technology, as it would provide for determining norms on location, design, construction, operation and maintenance (O&M) and decommissioning of the systems. This issue is critically important, because, in many situations, low-cost technologies or technologies preferred by local Panchayats for their ease of O&M are chosen over what can ensure sustainable supplies. It is also essential to understand the nature of the risks posed to water supply systems and how the 'degree of risk' changes from system to system and from locality to locality. This will help keep sufficient legal provisions in the proposed Act for risk aversion or reduction.

Moreover, for any water supply system to be sustainable, institutions play a crucial role. Currently, the institutions involved in drinking water supply management are only government agencies. Local community institutions, which are direct stakeholders, have a limited or no role. Merely transposing institutional models – private sector service providers, community institutions – and specific rules and regulations pertaining to the issuance of water connections, as well as penalties for non-payment of water charges and illegal tapping of water, that have worked elsewhere will not produce the desired results unless we understand the conditions that are conducive to the effectiveness of such institutions. The Act should take this into account.

The State Water Policy of Rajasthan (GOR, 2010) assigns the highest priority to drinking water supply for human consumption in water allocation decisions. However, there is a multiplicity of Acts enacted by various government departments that interface with water supply. There is a need to consolidate the provisions of these Acts to provide unified direction and interpretation. In this chapter, we present the outcomes of the review of a total of fifty-two Acts from fifteen states of India. Some of these Acts pertain to Municipalities/Municipal Corporations and Gram Panchayats, which also deal with water supply and sanitation as part of their mandate. Others are directly concerned with water supply and sewerage as framed by the states or local self-governing institutions.

2. A Review of Water Supply and Sewerage Acts in India

Before any new act is framed in any sector, it is important to examine the existing laws, acts, rules and regulations in that sector or in other sectors that have bearing on the said sector for the following reasons. First, we need to ensure that the new act doesn't violate the provisions of existing laws or Acts in that sector or the fundamental rights of citizens enshrined in the Constitution and that it is fully aligned with them (Cullet, 2016; Panickar, 2007; Upadhyay, 2011). Second, if the new act needs to supersede existing laws or Acts in any manner, it should be fully understood by those who frame the new Act or law. Third, a review of existing Acts in other parts of the country or in other countries, and the socioeconomic and political contexts in which they are embedded, would help those who draft the new Act identify the essential elements/ingredients of an Act, the frontiers that the new Act can explore, and also help them understand their performance on the ground, with insights into the difficulties in implementing them, especially the transaction costs.

Table 3.1 provides a list of the eighteen Municipal Council/ Municipality/Municipal Corporation Acts from fifteen states of India that were reviewed. Five Indian cities have framed their own Acts for water supply and sewerage. They are: the Hyderabad Metropolitan Water Supply Act (1989), the Delhi Water Board Act (1998), the Bangalore Water Supply and Sewerage Act (1964), and the Chennai City Municipal Corporation Act (1919). These were also reviewed. Furthermore, ten Indian states were found to have water supply and drainage/sewerage Acts. Thirteen states have Panchayati Raj Acts, which mandate the Gram Panchayats to provide water and sanitation services. These were also reviewed. Two states (Maharashtra and Uttar Pradesh) have *Zilla Parishads* Acts, with water supply coming under their purview. The states of Maharashtra and Karnataka have groundwater regulations for the protection of drinking water sources. A quick review of these Acts shows that barring the Delhi Common Effluent Treatment Act (2000), they have more or less common features.

It appears that most of these Water Supply Acts do not consider water resource availability as a constraint to meeting domestic water requirements. It also appears that most of the state water supply agencies or their pollution control boards are concerned with the quality of water at the source and not whether the same quality is maintained at the end-user level. In fact, most of the accepted water quality criteria for source protection are taken from the Environment Protection Act, 1986 (for acceptable quality of wastewater discharge into water bodies) or from various ISO standards, which are too broad to be effective in a regional context. The ecological carrying capacity of the 'sink' is a major determinant of the quality of effluent being discharged into it (Rajaram and Das, 2008). However, uniform norms are adopted across regions and states. The ecological carrying capacity changes remarkably from a 'hot and arid' region to a 'cold and humid' region, which means that the norms for effluent quality will have to be more stringent in a 'hot and arid region' compared to a 'cold and humid' region. None of the Acts have penal provisions to protect drinking water sources from contamination or to prevent illegal water tapping from the water supply system.

All the above legislative documents have been retrieved from the International Environmental Law Research Centre (IELRC) webpage on water law documents: <http://www.ielrc.org/water/doc_states.php>

3. A Review of Water Supply Legislation from around the World

Some of the important water service acts in other countries are presented in Table 3.2. The countries and regions included are: the United States of America, the European Union, Japan, England and Wales, Ireland, Malaysia, Ghana, Canada (specifically the province of British Columbia), South Africa and the Bahamas. The areas of concern in the water supply sector that are addressed by the legislation and their key features are discussed.

A quick review of these legislations reveals that their provisions are quite specific, dealing with critical areas within the drinking water sector such as the standards for water supply services regarding the quality and quantity of water to be supplied; the standards for the water supply infrastructure (facilities) to be created; drinking water protection and source protection; source siting, design and construction, O&M and decommissioning of water and wastewater treatment systems; drinking water distribution and wastewater collection; training and certification of operators; water treatment standards; monitoring, sampling and testing of water; specifications for the analysis of water quality parameters; collection, recording and reporting of information; supervision of water utilities; and emergency measures in response to the contamination of drinking water sources.

Table 3.1: Water Supply Legislations in Indian States

State	Legislation	Summary of Key Features
Andaman and Nicobar	The Andaman and Nicobar Islands (Panchayats) Regulation, 1994	In the sphere of sanitation and health, there is a provision for the maintenance of the drinking water supply.
Andhra Pradesh	Andhra Pradesh Municipalities Act, 1965	Part V, Chapter 1, covers the following with respect to water supply: vesting of works in councils; construction and maintenance of water works; constitution of water boards for local authorities; trespass on premises connected with water supply; prohibition of building over water mains; council to provide water for use; control over water supply connections; private water supply for consumption and use; power of the chairperson to enforce provision of water supply; power of council to make by-laws for water supply; levy and collection of pipeline service charges; and power to disconnect water supply.
	The Hyderabad Metropolitan Water Supply and Sewerage Act, 1989	This act has been framed for the provision of water supply, sewerage and sewage treatment in the Hyderabad metropolitan area and for matters connected therewith. Chapter II, Section 7 mentions that it shall be the duty of the board to supply potable water, including planning, design, construction, maintenance, operation and management of the water supply system, as well as sewerage, sewage disposal and sewage treatment works, including planning, design, construction, maintenance, operation and management of all sewerage and sewage treatment works in the Hyderabad metropolitan area.
Arunachal Pradesh	The Arunachal Pradesh Panchayat Raj Act, 1997	Schedule 1, Section VIII mentions the following duties of the Panchayat Raj Department with respect to drinking water: construction, repair and maintenance of drinking water wells, tanks and ponds; prevention and control of water pollution; and maintenance of rural water supply schemes.

(*continued*)

Table 3.1: Continued

State	Legislation	Summary of Key Features
Assam	Assam Municipal Act, 1956	In Chapter VII, Section 264, the following rules have been formulated with respect to the supply of drinking water: every Municipal Board shall provide or arrange for the provision of a sufficient supply of drinking water for the inhabitants of the areas within its jurisdiction; the Board shall make adequate provision for ensuring that the water supply is continuous throughout the year and that the water is at all times pure and fit for human consumption; and a Municipal Board shall also provide or arrange for the provision of a sufficient supply of water for other domestic or non-domestic purposes.
Bihar	Bihar Municipal Act, 2007	Chapter VI describes the core municipal functions. Every municipality shall provide on its own or arrange to provide through any agency the following core municipal services: water-supply for domestic, industrial, and commercial purposes; drainage and sewerage; solid waste management; preparation of plans for development and social justice; communication systems; construction and maintenance of roads, footpaths, pedestrian pathways, transportation terminals for both passengers and goods, bridges, over-bridges, subways, ferries, and inland water transport systems; transport system accessories including traffic engineering schemes, street furniture, street lighting, parking areas, and bus stops; community health and protection of the environment including planting and caring for trees on roadsides and elsewhere; markets and slaughterhouses; promotion of educational, sports and cultural activities; and the maintenance of an aesthetic environment. Furthermore, the municipality is expected to perform other such statutory or regulatory functions as may be provided by or under this Act or any other law in force at the time.

Chhattisgarh	Chhattisgarh Municipal Corporation Act, 1956	Under the Act, provisions were made for the supply of water.
	The Chhattisgarh *Peya Jal* Parirakshan Adhiniyam, 1986	The Act provides for the preservation of water in the water sources and the regulation of the digging of tube wells in order to maintain the public's water supply for domestic purposes and matters ancillary thereto.
Delhi	New Delhi Municipal Council Act, 1994	Water supply and sanitation provisions are provided under the Act.
	The Delhi Water Board Act 1998	This Act provides for the establishment of a board to discharge the functions of water supply, sewerage, and sewage disposal and drainage within the National Capital Territory of Delhi and for matters connected therewith.
	Delhi Common Effluent Treatment Plant Act, 2000	The Act provides for the recovery of the dues as arrears of land revenue in respect of the capital and recurring costs of common effluent treatment plants set up in the Industrial Estates in the National Capital Territory of Delhi and matters connected therewith or incidental thereto.

(*continued*)

Table 3.1: Continued

State	Legislation	Summary of Key Features
Goa	The Goa Panchayat Raj Act, 1994	Section 79 provides provisions for adequate water supply. For providing the area under its control or any part thereof with a supply of pure and sufficient water for public and private purposes, the Panchayat may: construct, repair and maintain tanks or wells and clear streams or watercourses; purchase or acquire by gift or otherwise any tank, well, stream or watercourse; or any right to take or convey water within or without the area under its control; with the consent of the owner thereof, utilize, cleanse or repair any tank, well, stream or watercourse, or provide facilities for obtaining water therefrom; contract with any person for the supply of water; or do any other act for carrying out the purpose of this section. The Panchayat may, by order published at such place as it may think fit, set aside for the supply of water to the public for drinking or culinary purposes any tank, well, stream or watercourse in respect of which action has been taken under clause (a), (b) or (c) of sub-section (1) subject to any rights that the owner referred to in clause (c) of sub-section (1) may retain with the consent of the Panchayat. The Panchayat may, by order published at such place as it may think fit, prohibit: bathing, washing of clothes and animals or other acts likely to pollute the water of any tank, well, stream or watercourse set apart for drinking or culinary purposes under sub-section (2); and the use of any source of water supply for drinking or culinary purposes or for the washing of clothes during epidemics.
Gujarat	Gujarat Water Supply and Sewerage Board Act, 1978	This Act provides for the establishment of a water supply and sewerage board for the rapid development and proper regulation of water supply and sewerage services in the state of Gujarat
Haryana	Haryana Panchayati Raj Act, 1994	Gram Panchayat has to take up the responsibility for: construction, repairs and maintenance of drinking water wells, tanks and ponds; the prevention and control of water pollution; and the maintenance of rural water supply schemes.

Himachal Pradesh	The Himachal Pradesh Panchayati Raj Act, 1994	This Act provides for water supply and sanitation
	Himachal Pradesh Water Supply Act, 1968	This Act provides for the development, control and management of the water supply works in rural and urban areas of Himachal Pradesh
Jharkhand	The Jharkhand Panchayat Raj Act, 2001	In Chapter VII on the functions of Panchayats, there is a provision for drinking water facilities.
Karnataka	Karnataka Urban Drinking Water and Sanitation Policy, 2003	This policy covers the provisions for water supply in the state of Karnataka.
	The Karnataka Municipal Corporations Act, 1976	Provisions for water supply include the water supply cess and the protection of the water supply premises.
	Karnataka Municipal Corporations (Water Supply) Rules, 2004	These rules were framed mainly for the construction and operation of water works through private parties.
	Karnataka Urban Water Supply and Drainage Board Act, 1973	This Act provides for the establishment of a Water Supply and Drainage Board and the regulation and development of drinking water and drainage facilities in the urban areas of the state of Karnataka.
	The Karnataka Ground Water (Regulation for Protection of Sources of Drinking Water) Act, 1999	Chapter II provides for protection measures for public sources of drinking water.
	Bangalore Water Supply and Sewerage Act, 1964	This Act provides for water supply, sewerage and sewage disposal in the Bangalore Metropolitan Area and for matters connected therewith.
	The Karnataka Panchayat Raj Act, 1993	Powers and duties with regard to sources of water supply are included in Chapter IV, Section 82.

(*continued*)

Table 3.1: Continued

State	Legislation	Summary of Key Features
Kerala	The Kerala Panchayat Raj Act, 1994	Water supply and sanitation provisions are provided under the Act.
	The Kerala Water Supply and Sewerage Act, 1986	This Act provides for the establishment of an autonomous authority for the development and regulation of water supply and wastewater collection and disposal in the state of Kerala and for matters connected therewith.
Madhya Pradesh	Madhya Pradesh Peya Jal Parirakshan Adhiniyam, 1986	This Act specifies the preservation of water in the water sources and the regulation of the digging of tube wells in order to maintain the water supply to the public for domestic purposes and for matters ancillary thereto
Maharashtra	The Maharashtra Groundwater (Regulation for Drinking Water Purposes) Act, 2013	This Act regulates the exploitation of groundwater for the protection of public drinking water sources and provides for matters connected therewith and incidental thereto
	Mumbai Municipal Corporation Act, 1988	Under this Act, a provision for water supply was made under Chapter X.
	The Maharashtra Zilla Parishads and Panchayat Samitis Act, 1961	The Act has provisions on water supply and sanitation by the Zilla Parishads.
Manipur	Manipur Municipalities Act, 1976	The Act provides for water supply and sanitation by the Municipalities.
Mizoram	The Mizoram Water Supplies (Control) Act, 2004	This Act provides for and regulates water supply to houses, land, buildings or premises belonging to any persons, government, central or state, or any organisation in the state of Mizoram and for other matters connected therewith

Nagaland	Nagaland Municipal Act, 2001	The Act provides for water supply and sanitation by the Municipalities.
Odisha	Odisha Municipal Corporation Act, 2003	Chapter XVII provides for the provision of water supply by the Municipal Corporations.
	Odisha Urban Sanitation Strategy, 2011	The overall goal of the strategy is to transform urban Odisha into community-driven, totally sanitized (safe), healthy and livable cities/towns that are managed by Urban Local Bodies with citizen and stakeholder participation.
	Orissa Municipal Act, 1950	The Act has provisions on water supply and sanitation by Municipalities.
	Orissa Water Supply and Sewerage Board Act, 1991	The Act has provision for the establishment of a Water Supply and Sewerage Board for the development and proper regulation of water supply and sewerage services in the state of Orissa and for matters incidental thereto or connected therewith.
Punjab	The Punjab Panchayati Raj Act, 1994	The Act has provisions for water supply and sanitation services by the Panchayats.
Rajasthan	Rajasthan Municipalities Act (as amended in 2010)	Provisions for water supply, drainage, sewerage and sewage disposal are included.
	The Rajasthan Panchayati Raj Act, 1994	Provisions for water supply and sanitation are included.

(*continued*)

Table 3.1: Continued

State	Legislation	Summary of Key Features
Tamil Nadu	Chennai City Municipal Corporation Act, 1919	Schedule XI includes provisions for water supply for domestic, industrial and commercial purposes.
	Tamil Nadu District Municipalities Act, 1920	Part IV, Chapter VII covers provisions for the public's health, safety and convenience with regard to water supply.
	The Tamil Nadu Panchayats Act, 1994	Schedule IV covers provisions for drinking water, health and sanitation, water management and watershed development.
	The Tamil Nadu Water Supply and Drainage Board Act, 1970	This Act provides for the establishment of the Water Supply and Drainage Board and the regulation and development of drinking water and drainage in the state of Tamil Nadu, except in the Madras Metropolitan Area.
Tripura	Tripura Panchayats Act, 1993	There are provisions for the establishment, repair and maintenance of rural water supply schemes; the prevention and control of water pollution; and the implementation of rural sanitation schemes.
Uttar Pradesh	Uttar Pradesh Municipalities Act, 1916	Sections 224–235 deal with the provisions of water supply and sanitation.
Uttarakhand	Uttar Pradesh Municipal Corporations Adhiniyam, 1959	In Chapter XI, there are provisions for water supply and the construction and maintenance of corporation waterworks.
	Uttar Pradesh Water Supply and Sewerage Act, 1975	This Act includes provisions for water supply in Chapter VII.
	U. P. Kshettra Samitis and Zilla Parishads Adhiniyam, 1961	This Act includes provisions for water supply and sanitation.

	The U. P. Panchayat Raj Act, 1947	Chapter IV details the powers, duties, functions and administration of the Gram Panchayat. There are provisions for drinking water, including the construction, repair and maintenance of public wells, tanks and ponds for the supply of water for drinking, washing and bathing purposes and the regulation of sources of water supply for drinking purposes.
West Bengal	West Bengal Municipal Corporation Act, 2006	Part IV, Chapter III mentions the powers and functions of the Corporations. There are provisions for the supply of water for public and private purposes.
	The West Bengal Municipal Act, 1993	In Chapter XV, under Section 226, there are provisions for water supply.
		Part V, Chapter XVII, Sections 259 and 260 provide the provisions for water supply.
	West Bengal Panchayati Raj Act, 1973	Chapter III, under Section 19, mentions the obligatory duties of the Gram Panchayat. There is a provision for the supply of drinking water and the cleansing and disinfecting of the sources of supply and storage of water.

Source: International Environmental Law Research Centre (no date).

Table 3.2: Some Water Supply Legislations from around the World

Country	Legislation	Summary of Key Features
USA	Safe Drinking Water Act, 1974–2004	The Safe Drinking Water Act (SDWA) was originally passed by Congress in 1974 to protect public health by regulating the nation's public drinking water supply. The law was amended in 1986 and 1996 and lists many actions to protect drinking water and its sources – rivers, lakes, reservoirs, springs and groundwater wells. However, SDWA does not regulate private wells that serve fewer than twenty-five individuals. SDWA authorizes the United States Environmental Protection Agency (US EPA) to set national health standards.
Ireland	Water Services Act, 2007 of Ireland	Part III solely provides for water supply, including misuse of water supplies, general duties of an owner, connection to water supply, conservation of water, pollution of water, taking samples of water, and the power to order the closure of a polluting water source.
European Union	European Union (Drinking Water) Regulations, 2014	In its schedule, the following four parts are covered: Part 1, parameters and parametric values; Part 2, monitoring; Part 3, specifications for the analysis of water quality parameters; Part 4, matters to be specified in the grant of departure under regulation 11.
Japan	Waterworks Act, 1957	This Act specifies drinking water quality standards; standards for facilities; application for a license of a water utility and its approval; operation of a water utility; supervision of water utilities; and the government's financial assistance and other rules.
The Bahamas	South Eleuthera Water Supply Act, 1962	This Act authorizes the agreement between the government and the South Eleuthera Power and Light Company Limited for the establishment of a water supply system in the settlement of Rock Sound and its environs and for matters incidental thereto.
England and Wales	Water Supply Regulations, 2010	This Act deals with the provisions of water supply and water quality.

Malaysia	Water Supply (Federal Territory of Kuala Lumpur) Act, 1998	This Act provides the law relating to the supply and distribution of water in the Federal Territory of Kuala Lumpur and for matters incidental thereto.
Ghana	Ghana Water and Sewerage Corporation Act, 1965	Under LI 1233, the corporation has enacted regulations through which water and sanitation facilities can be made available to Ghanaians. These regulations are meant to protect the environment, protect natural watercourses and provide for proper sewerage systems.
The State of British Columbia, Canada	A series of Regulations on Drinking Water; Wastewater, Sources of Drinking Water and Health Hazards The DWPA, SBC, 2001; The GWPA, 299/2004; The HHR, 216/2011; The MWAR, 87/2012	They cover the whole range of issues from drinking water source protection to location of sources, design and construction, operation and maintenance and decommissioning of water and wastewater treatment systems; drinking water distribution and wastewater collection; training and certification of operators; water treatment standards; monitoring, sampling and testing water; collection, recording and reporting of information; and emergency measures in response to contamination of drinking water sources.
South Africa	Water Services Act (2007)	The Act defines the 'rights of access to basic water supply' in volumetric terms; establishes norms and standards for water tariffs; prescribes who can be the water supply service providers; and contains legal provisions for the establishment of Water Boards. Similarly, the accountability on the part of the other stakeholders also needs to be fixed.

In contrast, the legislations in Indian states and Union Territories are largely meant to give powers to certain government and quasi-government bodies to undertake water supply and sewerage services and are quite general in nature. They have no clear provisions for setting the water supply service levels and water quality to be maintained by the utility; protecting the source and water supply from contamination; or deciding the price of water and penalties for non-payment of water charges. They also do not address the duties of the water utility and the consumers.

4. Existing Legislation for Regulating and Controlling Groundwater Over-Development

A large majority of the drinking water sources in India tap water from underground sources, which is also the major source of water to meet irrigation demands that account for significant portion of consumptive water use in India (Kumar *et al.*, 2022). For centuries, *de jure* rights to groundwater in India remained unclear. However, *de facto*, the right to use groundwater is attached to land ownership rights. Landowners can abstract the groundwater underlying their piece of land, and there are no restrictions on the volume of water that can be abstracted (Singh, 1995). This follows the English common law implemented during British rule[1]. Groundwater use and overuse can raise important equity issues, particularly in rural areas where increased consumption lowers the water table. In such situations, only the richer landowners can afford to sink deep wells and purchase the more powerful pumps necessary to abstract groundwater (Devi, 1991). Those with access to such water have a captive market comprised of those who do not. Equity issues also arise farming communities, cities, factories and other industrial sites that compete for groundwater against users in poorer rural areas (Hodgson, 2004). Hence, the nature of the legislation enacted in recent years to regulate the use of

1 Under English common law, any rights asserted to 'moveable and wandering' water must be based upon rights to the 'permanent and immovable' land below.

groundwater is extremely important in determining the sustainability of drinking water sources.

Several Indian states have formulated and passed legislation to deal with groundwater overexploitation. These include Maharashtra, Madhya Pradesh, Himachal Pradesh, Kerala, Tamil Nadu, West Bengal and Karnataka. The relevant acts are: Karnataka Groundwater (Regulation and Control of Development and Management) Act 2011 (GoK, 2011); Kerala Groundwater (Control and Regulation) Act 2002 (GoK, 2002); West Bengal Ground Water Resources (Management, Control and Regulation) Act, 2005 (SWID, 2005); Maharashtra Groundwater (Development and Management) Act 2009 (which came into force in 2013) (GoM, 2013); Tamil Nadu Groundwater (Development and Management) Act 2003 (Go TN, 2003a); AP Water, Land and Trees Act 2002 (GoAP, 2002); and Himachal Pradesh Ground Water (Regulation and Control of Development and Management) Act, 2005 (GoHP, 2005).

The legislations of Maharashtra and Karnataka specifically aim to regulate groundwater development to protect drinking water wells in 'notified-areas'. However, neither of them prescribes limits for groundwater abstraction by individual sectors such as agriculture, domestic, livestock and industry, nor suggests the volumetric water consumption by individual pumpers within each sector, which falls under the domain of the law governing 'water-rights'. They also do not provide for the creation of any institutional mechanisms for the allocation of water from common pool resources for domestic uses (especially during droughts) or its judicious use in order to meet the annual demand.

From a theoretical perspective, these legislations are not very comprehensive. They do not cover drinking water sources based on surface water bodies such as lakes, tanks and reservoirs, even though it is known that groundwater and surface water systems are hydraulically interconnected, with water abstraction from aquifers affecting inflows and storage in surface water bodies and vice versa. Hence, these regulatory measures are largely ambiguous in terms of their intended outcomes.

In terms of actions, some of the legislation suggests banning new wells in a 'notified area'. Others prescribe a minimum distance between wells in a 'notified area', and some specify the distance to be maintained by a newly constructed well meant for non-drinking purposes from a drinking water source maintained by a state-run water supply agency. The AP Water, Land and Trees Act – 2002, for instance, makes registration of all wells and permission for sinking new irrigation wells near drinking water sources mandatory with the newly constituted Water, Land and Trees Authority. It also prohibits water pumping in certain 'notified areas' and provides for compensation to the owners of wells that are closed for the protection of drinking water sources based on groundwater (*APWALTA 2002*).

Questions can be raised about the feasibility of enforcing such regulations, especially in remote rural areas, given the current administrative setup of the agencies responsible for their enforcement. In the context of the Maharashtra Groundwater (Regulation for Drinking Purpose) Act 1993, Phansalkar and Kher (2006), while raising doubts about the ability of Gram Panchayats to take action against villagers who violate the Act, also questioned the validity of such classifications, as the agencies that are responsible for groundwater assessment and notification of 'over-exploited areas' can be subject to significant pressure from local groups to classify an area as 'safe' for groundwater development.

Regarding the effectiveness of these actions, banning new wells and enforcing well spacing regulations may not result in reduced groundwater draft but will have serious negative equity impacts, as many new well drillers in over-exploited areas are likely to be less privileged farmers. Hence, such measures, if attempted, will have serious social and political ramifications. Also, prescribing a uniform minimum spacing between wells to prevent well interference without considering the geo-hydrological environment doesn't make hydrological sense. In hard rock areas, the distance to be maintained between wells to avoid interference would be much larger than in an alluvial area due to differences in aquifer yield characteristics.

The groundwater regulations attempted in Indian states so far, therefore, are neither hydrologically meaningful nor socially viable. By merely putting restrictions on the number of new wells with no restrictions on the energy use for pumping groundwater through existing wells, groundwater cannot be conserved for future use, notwithstanding the ability of the regulatory authority to monitor well drilling activities in rural areas. Any Act proposed for Rajasthan to regulate groundwater development for the protection of drinking water sources has to take into account that the state has heterogeneous geo-hydrology and that farmers use both electric pumps and diesel wells to pump groundwater.

5. Summary and Conclusions

The extensive review of drinking water acts suggests that the Acts that the existing acts in Indian states and local bodies (such as Corporations, Municipalities and Gram Panchayats) are primarily designed to create bodies responsible for the development of drinking water sources, sewerage and drainage facilities, and the regulation of water supply services (Panickar, 2007). These Acts specify the constitution, powers, duties, and key functions of these bodies. Very few Acts provide for the protection of water supply or source water. None of them specify the water quality standards for the supplied water. The state-level water supply agencies and the local water utilities generally follow the guidelines specified in the manuals of the Central Public Health and Environmental Engineering Research Institute (CPHEERI) or the specifications of the Bureau of Indian Standards (BIS) for determining drinking water quality requirements and service levels. They are not mandated by any law to strictly adhere to these standards or guidelines.

The legislative measures to control groundwater development adopted by the governments of various Indian states (e.g. Maharashtra, Karnataka, Madhya Pradesh, Kerala, West Bengal, Goa, etc.) are largely ambiguous in terms of their intended impacts, hydrologically meaningless and socially unviable. The institutional capacities of the agencies mandated to enforce the legislation and catch the 'offenders', who are often in remote rural areas, are also highly questionable.

In contrast to the situation in India, where the drinking water Acts are quite basic and simple, those in many developed countries are far more advanced and mature. They contain provisions to deal with the complex challenges faced by the sector. These acts specify the standards for drinking water quality (EU, England and Wales, US, Japan and the state of British

Columbia, Canada), standards for the water supply facilities to be created (Japan), monitoring of water supply for quality and protection of public health (US and EU) and protection of drinking water supply and source waters (US, Japan, EU and the state of British Columbia, Canada). They also specify standards for monitoring, sampling and testing water for quality checks; and emergency measures in response to the contamination of drinking water sources (the state of British Columbia, Canada). The Water Services Act of South Africa defines the right of access to water for basic needs as a fundamental right.

CHAPTER 4

A Drinking Water Supply Services Act for Rajasthan

1. Introduction

The guidelines issued by the National Rural Drinking Water Programme (NRDWP) (2013) under the Ministry of Drinking Water and Sanitation of the Union Government to ensure household water security and long-term sustainability of the services, which were meant to provide a framework for implementation by the state governments, are conspicuous by the absence of any suggestions to the state governments on the need for legal instruments and institutional measures to achieve drinking water security. This is in spite of the fact that no Indian state has come up with legislation on drinking water supply services that addresses the concerns of both rural and urban areas and is comprehensive enough to address all the concerns (Panickar, 2007). These concerns are related to the physical, social, financial, economic, environmental and institutional sustainability of the services. Instead, these guidelines largely pontificate on the need for adopting measures such as rainwater storage, the use of multiple sources of water, protecting water sources, groundwater recharging and the use of regional pipelines (NRDWP, 2013). In this chapter, we present the draft Act on drinking water in Rajasthan in the form of chapters and articles. There are a total of eighteen chapters and seventy articles. Before presenting the Act, a preamble is provided, followed by various terms and definitions used for framing the Act included in Chapter 1. Chapter 2 presents the general provisions of the Act, which are followed by various articles (seventy in total) presented in the remaining sixteen chapters.

The thematic areas dealt with in the remaining sixteen chapters are: (1) treatment of raw water for domestic water supply; (2) norms on quantity, quality and frequency of drinking water supply by public utilities; (3) issuing of water connections; (4) treatment and disposal of wastewater; (5) fixing of water tariff; (6) selection of water supply technologies; (7) location, design and construction of water supply systems; (8) operation and maintenance of water supply systems; (9) recycling and reuse of wastewater; (10) water supply surveillance; (11) functioning of water service providers; (12) illegal water tapping, drinking water monitoring, protection of drinking water sources and contamination & pollution of 'drinking water sources'; (13) institutional capacity building for water supply; (14) guidelines and criteria for convergence of government programmes in rural areas; (15) agency for bulk water supply; and (16) water supply, sanitation and hygiene.

2. The Preamble

Access to an adequate quantity and quality of water is necessary for human survival and improved health. Rajasthan is one of the most water-scarce regions in the world. Provisioning water of adequate quality for human consumption is a major challenge for the Public Water Utilities in both rural and urban areas of the state. The 'Drinking Water Supply Services Act for Rajasthan', hereafter referred to in this document as 'the Act', is aimed at facilitating effective and efficient management of drinking water for promoting public health and hygiene. It will also contribute to the needs of all consumers of water, Public Water Utilities, water supply institutions, water supply service providers and all agencies concerned with the protection and management of water resources in the state of Rajasthan. This Act provides a set of standards, criteria and guidelines pertaining to all aspects of drinking water supply and utilization, wastewater treatment and disposal, and recycling and reuse of treated wastewater in rural and urban areas of the state. The rules and regulations will be formulated after the draft Bill is passed by the Legislative Assembly and becomes an Act.

3. The Drinking Water Services Act

3.1. Chapter 1: Terms and Definitions

3.1.1. Access to Adequate Quantity and Quality of Water

3.1.1.1. QUANTITY

The per capita water supply for domestic uses, to be maintained by government schemes for rural areas and different types of urban areas, shall be as per the CPHEEO Manual (May 1999), Ministry of Urban Development for urban areas and NRDWP guidelines (August 2013) of the Ministry of Rural Development for rural areas, with future revisions therein from time to time.

3.1.1.2. QUALITY

Quality of water refers to the physical, chemical, microbial and radiological content of the water. Water of adequate quality is water that is either free from these impurities or has these impurities within permissible limits for the 'specified use', as defined by the standards followed by the state of Rajasthan.

3.1.2. Basic Survival Needs for Water

As per NRDWP guidelines, 8 litres per capita per day should be treated as a basic water requirement for drinking and cooking purposes.

3.1.3. Bulk Water Supply Agencies

Bulk water supply agencies are those that are mandated to and have the powers, under this Act, to allocate water from surface water sources on a volumetric basis for use among various 'water rights holders'. Such water rights holders, as recognized by this Act, are Public Water Utilities engaged in any of the following: domestic water supplies in rural areas, drinking water supplies in urban areas, and/or industrial water supply. The surface water sources shall include: man-made reservoirs, lakes; rivers/streams, and tanks.

3.1.4. Capacity Building

Capacity building is the process of gaining technical, managerial, and institutional knowledge and insight in relation to the socio-economic structure, cultural standards and values of the society concerned. It also involves skill development in technical spheres.

3.1.5. Communication Pipe

A communication pipe is any pipe or system of pipes, along with all fittings thereto, by means of which water is supplied to any user, licensee, premises, habitation, village or area from the main pipe and includes a connection pipe, service pipe, meter or other fittings.

3.1.6. Community-Based Organizations (CBOs)

A community-based organization is representative of a community or a significant segment of a community and is engaged in meeting its human, educational, environmental or public safety needs. It can either be public or a private non-profit.

3.1.7. Connection Pipe

It means any water pipe from a ferrule to a stopcock connecting the main with the service pipe.

3.1.8. Consumer

Any person getting the benefit of the water supply provided by the Public Water Utility

3.1.9. Domestic Wastewater

Domestic wastewater is the waste carried through water, in solution or suspension, from dwellings or from a community. It is mostly water and characterized by volume, physical condition, chemical and toxic constituents, and bacteriologic status. It consists mostly of greywater (from sinks, tubs, showers, dishwashers, and clothes washers); blackwater (from flush toilets, combined with the human waste that it flushes); soaps and detergents; and toilet paper.

3.1.10. Drinking Water Contamination

Water contamination may be natural or man-made. It refers to impurities in drinking water regardless of the source or events that lead to the impurities being in the drinking water.

3.1.11. Effluent Quality Guidelines

Guidelines are available from WHO 1989 for the use of treated wastewater in irrigation, concerning human health protection and crop production. CPCB guidelines for discharge in natural drains are known as effluent quality guidelines.

3.1.12. Equity

Equity is defined in a broad sense here, covering distributional or spatial equity, social equity, gender equity and intra-generational equity.

3.1.13. Groundwater

Groundwater is water that is held underground in the soil or in pores and crevices in rock

3.1.14. Human Resources Development (HRD)

HRD is the process of providing technical assistance to develop the capacity of local institutions such as the Municipality/Public Water Utility/PRIs for a variety of administrative, financial and managerial activities to perform water management functions effectively.

3.1.15. Illegal Tapping of Water

Illegal tapping of water refers to drawing water from a public water supply system without obtaining a connection through a formal application and without making payment for the same.

3.1.16. Institutional Reform

Institutional reform is the process of reviewing and restructuring institutions so that organisational goals are achieved in a better manner, preserving the rule of law and ensuring accountability to beneficiaries.

3.1.17. Main

A main is a pipe laid by the Public Water Utility for the purpose of providing a general supply of water as distinct from a supply to individual consumers. It includes any apparatus used in connection with such a pipe.

3.1.18. Occupier

An occupier includes an occupier of land, premises or other property

3.1.19. Owner

An owner is every person having ownership of land, premises or other property, along with all rights and obligations that attach to an owner under the provisions of the Act

3.1.20. Panchayati Raj Institutions

The Panchayati Raj Institutions (PRIs) are institutions of local self-governance in India, set up as per the Indian Constitution. They are part of the distributed governance system in India, operating at the levels of districts, blocks/talukas and village Panchayats. A village-level Panchayat often has jurisdiction over more than one village. Though they are institutions of governance, several managerial functions, which were earlier performed by various line departments and agencies of the government, are now being looked after by the Panchayats.

3.1.21. Premises

Premises means any building or part of a building or outhouse but shall not include a garden, ground or lawn, if any, for the purpose of inspecting a water connection.

3.1.22. Prescribed

Prescribed means prescribed by rules made under the Act.

3.1.23. Primary Treatment

Primary treatment involves basic processes to remove suspended solid waste and reduce biochemical oxygen demand in wastewater.

3.1.24. Private Sector Participation

Private sector participation in this Act refers to the participation of private agencies with regard to capital investment and/or skill development of a Public Water Utility for the execution of a new water supply scheme; the operation and maintenance of existing or new schemes; or any other designated area as decided by the Government of Rajasthan.

3.1.25. Rainwater Harvesting

Rainwater harvesting refers to the collection (accumulation) and storage (deposition) of rainwater at the point where it falls, before it runs off the area, for direct use for domestic, livestock, irrigation and industrial uses. In certain situations, the collected rainwater is diverted for deep percolation into underground formations.

3.1.26. Rooftop Rainwater Harvesting Systems

Rooftop rainwater harvesting refers to the process of harvesting the runoff generated from the precipitation falling on the roofs of buildings (residential, commercial or industrial) and either storing it in tanks for direct use or storing it in aquifers for future use after proper treatment.

3.1.27. Safe Water

Safe drinking water is water with microbial, chemical, radiological and physical characteristics that meet national standards for drinking water quality.

3.1.28. Secondary Treatment

Secondary treatement is a process that uses a sewage treatment plant with physical phase separation to remove settleable solids and a biological process to remove dissolved and suspended organic compounds to achieve a certain degree of effluent quality.

3.1.29. Service Pipe

Service pipe means any pipe other than the connection pipe beyond the stopcock by means of which water is supplied to any consumer or premises.

3.1.30. Sewerage System

The sewerage system is the infrastructure built to collect and transport sewage from residential areas and commercial establishments to a central point to be disposed of into a natural sink with treatment.

3.1.31. Surface Water

Surface water is water that collects on the surface of the ground, such as in a stream, river, lake, wetland or ocean

3.1.32. Tap Water

Tap water is the water that is supplied through a proper distribution system after the raw water is treated for potability and made 'safe' for human consumption. Any water that is available through pipelines and taps is not tap water, unless it is treated to safety standards.

3.1.33. Treated Wastewater

Treated wastewater is the wastewater treated to a specified standard with respect to physical, chemical and microbial quality using various treatment systems and processes. Treated wastewater can be classified into primary treated, secondary treated and tertiary treated.

3.1.34. Tertiary Treatment

Tertiary treatment is the highest level of treatment, and therefore the water from tertiary treatment would generally be potable.

3.1.35. Treated Water

Treated water is raw water available from natural systems – rainwater, surface runoff, spring water and groundwater – that is treated to remove physical, chemical and microbial impurities and meets the drinking water standards as per the norms and guidelines stipulated from time to time by competent authorities.

3.1.36. Wastewater

Wastewater includes domestic sewage and wastewater from manufacturing.

3.1.37. Wastewater Treatment Plant

A waste water treatment plant is a technical system designed as per standard design criteria to treat a designated volume of waste generated from residential areas, livestock farms, industrial units (small or large) or commercial establishments (hotels, restaurants, utility buildings) or some or all of them to a specified standard of physical, chemical and microbial content using physical, chemical or biological processes. The treatments are of four types: preliminary (removal of grit), primary (suspended sediments), secondary (biological process to reduce BOD) and tertiary (to remove nitrates, phosphates, heavy metals and microbes).

3.1.38. Water Connection

A water connection includes any tank, cistern, hydrant, stand-pipe, meter or tap situated on any premises or property and connected with a main or distribution pipe.

3.1.39. Water Metering

Water metering is the process of installing water meters in the water supply pipeline done by Utilities to check for leakages (technical losses) in the distribution system and to measure the volume of water supplied to consumers.

3.1.40. Water Pollution

Water pollution means the introduction into a water body of any substance that alters the chemical, physical, microbial or radioactive properties of the water in such a way as to affect its effective use, endanger human health, damage the ecosystem or be deleterious to water quality.

3.1.41. Water Pricing

Water pricing is a way to recover the cost of water supply. It is also a way to make consumers realize the scarcity value of the resource, thereby achieving efficient allocation of the resource. Efficient pricing of water plays an important role in regulating water use, ensuring equitable water distribution and improving the financial workings of the water utility.

3.1.42. Water-related Services

Water-related services mean water supply services and sanitation services.

3.1.43. Water Safety Plan

A Water Safety Plan (WSP) consistently ensures the safety and acceptability of a drinking-water supply. This is done by eliminating or minimizing potential risks of contamination in raw water sources, water treatment plants, catchments, distribution networks, storage, collection and handling.

3.1.44. Water Supply Institutions, Public Water Utilities, Water Service Provider

(i) A water supply institution refers to a Public Water Utility, a water service provider, a Village Water and Sanitation Committee (VWSC) or a Panchayati Raj Institution (PRI) for water-related services.
(ii) A Public Water Utility is a department of the state, a Municipality/Corporation, or an arm of the Municipality/Corporation established to perform a public function of water-related services.

(iii) A water service provider refers to any person or agency that provides water-related services to consumers or to another water services institution.

3.1.45. Water Supply Services

Water supply services mean the abstraction, conveyance, treatment and distribution of potable water for domestic consumption and commercial and industrial use.

3.1.46. Water Supply Surveillance

Water supply surveillance is an exploratory action intended to identify and evaluate factors associated with drinking water that could pose a serious health risk, aimed at improving the quality, quantity, coverage, cost and continuity of water supplies.

3.1.47. Water Supply System

A water supply system is a technical system used for acquiring water from a natural system, storing and treating water (if required), and supplying it in the desired quantity, quality and frequency to a single household, a few households, a community, an entire village/hamlet, a town, a city, or a group of villages, towns and cities.

3.1.48. Water Supply Technologies

Water supply technologies are any technologies used for extraction, treatment, pumping, conveyance and distribution of water from the natural system (including sea, river systems and sub-surface systems) to meet the water supply requirements of drinking water, domestic water, commercial water and industrial water users in terms of quantity and quality, either together or individually.

3.1.49. Water Tankers

Water tankers are tankers used for transporting water, generally to meet water needs in dwellings, commercial buildings, construction sites and industrial units. They are usually used by both public utilities and private water supply providers to meet water requirements.

3.1.50. Water Theft

See the definition of illegal water tapping.

3.1.51. Water Works

Water works are a system consisting of whole or part of various components such as a reservoir, dam, well, pump house, borehole, pumping installation, water treatment works, wastewater treatment plant, access road, electricity transmission line, pipeline, meter, fitting or apparatus built, installed or used by a water services institution to provide water-related services.

3.2. Chapter 2: General Provisions

Article 1: *Purpose of formulating the drinking water supply services regulations Act*

This Act provides a set of standards, criteria and guidelines pertaining to all aspects of drinking water supply and utilization in rural and urban areas of Rajasthan. This Act would facilitate effective and efficient management of drinking water, contributing to the needs of all consumers of water, Public Water Utilities, water supply institutions, water supply service providers and all agencies concerned with water quality protection and the management of water resources in the state of Rajasthan.

Article 2: *Water supply as an emergency service*

1. Water supply services shall be treated as an emergency service, and the officials of the Public Water Utilities shall have suitable powers to deal with the situation as described in the rules.

Article 3: *Who will the Act apply to?*

1. The various provisions of the Act shall apply to the Public Water Utilities, water supply institutions, water supply service providers, NGOs, community-based organizations and consumers, unless otherwise specified in the various articles.

Article 4: *Who will implement the provisions of the Act?*

1. The implementation of various provisions of the Act concerning water connections, treatment of raw water for public water supply, norms for quality, quantity and frequency of drinking water supplies, water quality management, water metering and water tariff, water supply technologies, location, design and construction of water supply systems, operation and maintenance of water supply systems, water supply surveillance, illegal water tapping, protection of drinking water sources, institutional capacity building, bulk water supply, and water sanitation and hygiene (provisions under Chapters 3, 4, 5, 7, 8, 9, 10, 12, 14, 15, 17, 18 and the various articles therein) shall be within the purview of the concerned Public Water Utilities.
2. The implementation of various provisions of the Act concerning the treatment and disposal of wastewater and recycling and reuse of wastewater (provisions under Chapters 6 and 11 and the various articles therein) shall be within the purview of the State Pollution Control Board of Rajasthan.

3. The implementation of various provisions of the Act relating to the functioning of water supply service providers (provisions under Chapter 13 and various articles therein) will be the responsibility of the Public Water Utility.

Article 5: *Who will enforce the provisions of the Act against violators?*

1. The agencies responsible for enforcing relevant provisions of the Act against violators are: the Public Water Utilities and law enforcement agencies in the case of violations concerning illegal water tapping and contamination and pollution of water source (Articles 49 & 50); the State Pollution Control Board in the case of violations concerning wastewater treatment and disposal and recycling and reuse of wastewater (Articles 20, 21, 22, 36, 37 & 38); the Public Water Utility in the case of violations concerning the functioning of water supply service providers (Articles 42, 43, 44 & 45), and monitoring and protection of drinking water sources (Articles 46 & 47).

Article 6: *Who will have the right to claim damages arising out of violations of the Act?*

1. The Public Water Utility will have the right to claim damages in case of any violations of provisions of the Act (under Articles 15, 46, 47, 48, 49 and Clause 2 of Article 50) by domestic, commercial, industrial and institutional users of water.
2. The nature and extent of damages will be assessed by the Public Water Utility as per their prescribed standards.
3. Consumers shall have the right to redress grievances with Public Water Utilities. If the consumer is not satisfied, the consumer court may be approached.

3.3. *Chapter 3: Treatment of Raw Water for Public Water Supply*

Article 7: *Standards for treatment of raw water for public water supply*

1. It shall be the duty and responsibility of the Public Water Utility to supply water of a certain standard, as per the CPHEEO manual and accepted by the Ministry from time to time, to the consumers for the purposes for which the said utility is mandated to perform.
2. The Utility shall treat the raw water obtained from natural systems used for public water supply to make it 'potable' from the point of view of physical, chemical, microbial and radiological properties, as per the standards accepted by the Ministry for Drinking Water, Government of India, before it is supplied through the system, if the water supply serves the needs of domestic consumers along with others.

Article 8: *Methods for treatment of raw water for public water supply*

1. The raw water to be used for treatment prior to public water supply shall be of such quality, as long as it is acquired from a natural source of water and conforms to drinking water quality standards after treatment.
2. The Public Water Utility shall select the method of treatment of the raw water, depending on its quality and its end use, to conform to drinking water quality standards.
3. The Utility may adopt sophisticated treatment methods to obtain the best quality, keeping in mind the life cycle cost and social acceptability. Such methods can include any of the following:
   (a) Sedimentation tanks for the removal of suspended matter (preliminary treatment)
   (b) Simple graded filters to remove physical contaminants and microbes
   (c) Filters of various types
   (d) Ion exchange process to remove bicarbonates
   (e) Defluoridation to remove fluorides

(f) Reverse osmosis
(g) Electrolysis
(h) Chlorination for disinfection
(i) Ozone treatment for disinfection
(j) Any other method approved by the Bureau of Indian Standards

3.4. *Chapter 4: 'Drinking Water Supplies' by Public Utilities: Norms for Quality, Quantity and Frequency*

Article 9: *Different water use sectors to be served by Public Water Utilities and Water Supply Institutions*

1. The water supplied by public water utilities and other water supply institutions in urban areas shall cater to the requirements of domestic users, commercial users, industrial water users and environmental management in terms of quantity, quality and frequency.
2. The water supplied in rural areas shall meet the requirements of the domestic sector, including water for direct consumption by domesticated livestock.

Article 10: *Norms and standards regarding the physical, chemical and microbial quality of drinking water*

1. The quality of water supplied by Public Water Utilities in both rural and urban areas should conform to the drinking water quality standards prescribed by the CPHEEO manual with respect to the physical, chemical and microbial quality of water.
2. The quality of 'water supplied to the consumers' shall be that of the water that is received at the end-user level and not that supplied from the system at the source of supply.
3. It is understood that the Public Water Utility will undertake measures to prevent contamination of water in the distribution and delivery systems.

Article 11: *Standards and norms regarding the quantity of drinking water*

1. The quantity of water, which is mandatory for the Public Water Utilities and other water supply institutions to supply to domestic consumers in per capita terms in both rural and urban areas, shall be the water received at the end-user level and not the water supplied from the system at the source.
2. The minimum quantity of water to be supplied in urban and rural areas shall be as per the new guidelines issued by the CPHEEO and NRDWP, respectively, with upward revisions, if required, as per the guidelines issued by the concerned Ministry at the Union level from time to time.

Article 12: *Norms regarding the frequency of drinking water supply*

1. The frequency of water supply from Public Water Utilities and other water supply institutions shall be 24 × 7, preferably in urban and rural areas, and it should not be less than once a day for a minimum duration of one hour in every season if continuous water supply is not maintainable.
2. The schedule for water supply to be maintained by the Public Water Utility and other water institutions in urban areas shall be known to the consumers in every locality.
3. Prior information on any changes in supply schedules shall be given to consumers through various communication media at least twenty-four hours in advance.
4. The water supply system for new areas shall be designed for continuous water supply to avoid the requirement for intermediate storage systems by consumers, and to reduce contamination of water in the pipes in the wake of intermittent supply.
5. Low-priority uses of water, such as the watering of lawns and gardens and the use of water in parks, shall be banned during crises, such as drought conditions.
6. During emergencies and crises, the Public Water Utility shall have the power to acquire any private water source in the state

in the public interest on payment of suitable rent, and the same shall be decided as per the rules framed by the department for the purpose.

7. Such acquisitions shall be valid only until the time period of the crisis is over.

3.5. *Chapter 5: Release of Water Connections*

Article 13: *Application for water supply connection*

1. The owner or occupier of any premises may apply to the Public Water Utility for the supply of water for domestic, non-domestic or industrial purposes in the prescribed form and accompanied by the prescribed fee.
2. The concerned official of the Public Water Utility shall review each application and, within thirty days after receipt of the application, either grant the water supply connection to the applicant or, for reasons to be recorded in writing, reject the request for such connection.
3. No person shall, without the authorization or permission of the Public Water Utility, make or cause to be made, at any time, any connection with a main line in contravention of the provisions of the Act.

Article 14: *Expenses related to water supply connections*

1. Provided that the request of the applicant for the grant of a water supply connection is accepted by the Public Water Utility, the applicant should, at his/her own cost, provide all communication pipes and fittings and create conditions to carry out works relating to the laying and jointing of such communication pipes and fittings in the premises for which the water supply is sanctioned.
2. If the person to whom such permission is granted fails to get his/her premises fitted with pipes and a necessary connection within

a period of two months from the date of the grant of permission, such permission shall be deemed to be inoperative.

3. In case a service line supplies water to two or more premises or consumers from the main line or distribution pipe, the Public Water Utility may by written notice, require the owners of such premises or such consumers to lay down separate branch pipes for such premises or such consumers, and the expenses of doing so shall be borne by all such owners or consumers in such proportion as may be determined by the Public Water Utility.

Article 15: *Prohibitions on certain acts*

1. No person shall unlawfully obstruct the flow of, flush, draw off, divert, or take water from any main or distribution pipe or any service line, tap work or fitting connected with the supply of water to the premises by using, fitting, installing or connecting therewith any pump or by using any other means.
2. The owner of a premise on which water supplied under the Act is misused shall be bound to prevent such misuse or inform the Public Water Utility about such misuse.

Article 16: *Power to enter premises and inspect*

1. Any concerned officer or gazetted officer authorized in this behalf by the Public Water Utility may enter any premises for the purpose of inspecting any water installation in any building, site or premises after informing the owner of the said premises.
2. If such officer is, at any such time, refused admittance into such building, site or premises for the purpose aforesaid or is prevented from making such inspection, the Public Water Utility may, after giving the owner of such building, site or premises an opportunity to be heard, cut off the supply of water to such building, site or premises.
3. The provisions of the *Code of Criminal Procedure, 1973*, shall, so far as may be, apply to any search or seizure under this section as

they apply to any search or seizure made under the authority of a warrant issued under the said Code.

Article 17: *Water supply disconnection*

1. Notwithstanding anything contained in the Act, the Public Water Utility may cut off the supply of water to any premises or may turn off such supply on any one of the following grounds:
   (a) if the owner or occupier of the premises continues to use the water or permits the same to be used in contravention of the Act or the rules and regulations framed thereunder, after the receipt of a written notice from the Utility requiring him/her to refrain from so doing;
   (b) if any pipe, tap or fitting connected with the supply of water to the premises is found, on inspection in this behalf, to be damaged or out of repair to such an extent as to cause waste or contamination of water, making immediate prevention necessary;
   (c) if there is any water pipe situated within the premises to which no tap or other efficient means of turning off the water supply is attached;
   (d) if, by reason of a leakage in the pipe or fitting, damage is caused to the public street and immediate prevention is necessary; or
   (e) if the owner of the premises refuses or fails to pay, for a period of one month after he/she has been served with a notice of demand, water charges on account of water consumed in the premises.
2. When the water supply has been cut off or turned off, the officer concerned may, on such terms and conditions as may be prescribed, recommend the restoration of the water supply upon payment of water usage charges and penalties as applicable.

Article 18: *Appeals*

1. An appeal can be made against the order concerning the disconnection of water supply to the next higher official of the Public Water Utility.
2. Every such appeal shall be preferred within ninety days from the date of the order for disconnection.
3. On receipt of any such appeal, the Public Water Utility shall, after giving the appellant a reasonable opportunity of being heard and after making such inquiry as it deems proper, dispose of the appeal within a period of fifteen days from the date of its institution.
4. The decision of the Public Water Utility on such an appeal shall be final.

Article 19: *Indemnity*

1. Notwithstanding anything contained in any law for the time being in force, the Government shall not be liable for any damage or penalty for discontinuing the supply of water or failing to supply water to any person, any premises, or any local area if the cause of such failure is beyond the control of the Government.

3.6. Chapter 6: Treatment and Disposal of Wastewater

Article 20: *Criteria for treatment of domestic wastewater*

1. This article concerns ensuring the protection of water bodies and drinking water sources and the treatment of wastewater to safe standards for reuse and disposal.
2. The minimum treatment for domestic wastewater from urban areas, including metros, cities and towns, is 'secondary treatment', as defined by the Water Pollution (Prevention and Control) Act 1974 of the Central Pollution Control Board.
3. Domestic wastewater from urban dwellings located in sparsely populated areas, where sewerage systems would be economically

unviable, should be treated with safe on-site sanitation technologies such as decentralized wastewater treatment systems, ensuring no threat to public health.

4. Domestic wastewater from all other urban areas should be treated to water quality standards safe for use in irrigation if the treated wastewater is discharged into artificial channels for direct use by irrigators.
5. Domestic wastewater from urban areas, which is to be discharged into natural watercourses, should be treated to standards suitable for bathing and the propagation of wildlife and fish populations prior to discharge.
6. Domestic wastewater from urban areas, treated using soil media and entering groundwater, should meet physical, chemical and microbial quality standards for drinking water.
7. Treated wastewater discharged into water bodies used as drinking water sources by the PHED and local Municipalities and Corporations, should meet acceptable water quality standards for drinking water sources.
8. Domestic wastewater from rural areas should be diverted into 'soak pits', prepared according to design criteria provided by the State Public Health Engineering Department.
9. 'Soak pits' can be constructed by individual households, groups of households or the community.
10. Funding for this can be sourced from the Swachh Bharat Mission or MNREGS through the concerned village Panchayat.
11. Supervision of this work shall be conducted by the Panchayat sub-committee on water supply and sanitation.

Article 21: *Standards for treatment of domestic wastewater*

1. These standards concern the quality of wastewater collected through the sewerage system and treated in centralized treatment plants.

2. Treatment standards for domestic wastewater should be determined by the point of discharge and type of reuse of the treated wastewater.
3. Domestic wastewater reused for irrigation should meet water quality standards stipulated by the effluent quality guidelines for human health and increased crop production of the World Health Organization (1989).
4. Water quality standards for wastewater discharged into natural watercourses should meet bathing standards for humans and standards for the propagation of fish and wildlife, as stipulated by the CPCB.

Article 22: *Guidelines for disposal of treated wastewater*

1. In urban areas, domestic wastewater treated to irrigation quality standards should be discharged into artificial drains supplying water directly to agriculture and should not contaminate natural watercourses.
2. Wastewater treated using soil media should not be disposed of in areas where it can percolate into freshwater aquifers currently used or potentially usable for drinking water supply in the future.
3. In such cases, percolating water should be pumped out for irrigation and industrial water supply before natural mixing occurs.
4. Tertiary-treated wastewater should not be discharged into natural drainage systems but should be reused for domestic water supplies with public perception in mind or discharged into water bodies serving as drinking water sources.
5. For soak pits, the location should ensure the water table is at least 10 metres below ground level.

3.7. Chapter 7: Fixing Water Tariffs

Article 23: *Criteria for fixing water tariffs in urban areas*

1. The power to fix water tariffs shall lie within the purview of the Public Water Utility.
2. Tariffs for water supplied by Public Water Utilities in urban areas shall be based on the purpose of water use (social production, economic production or environmental use), economic efficiency, cost recovery, and affordability.
3. Water tariffs shall be determined based on volumetric consumption for all uses, with differential tariffs applied for different levels of consumption using incremental block rates.
4. The tariffs set should promote water use efficiency and conservation.

Article 24: *Criteria for fixing water tariffs in rural areas*

1. Tariffs for drinking water supplied by Public Water Utilities in rural areas shall primarily consider cost recovery and affordability.
2. Introduction of pricing based on household consumption in rural areas shall be contingent upon economic feasibility.
3. Differential fixed charges shall be imposed on households obtaining domestic water connections, with or without metering, based on the economic status of the households, and minimum charges will be levied upon from BPL families.
4. For households with individual household-level metering is practised, differential, varying charges shall be recovered for apply based on different levels of consumption.
5. Unmetered connections in dwellings shall incur fixed monthly water charges.

Article 25: *Norms for fixing water tariffs in urban areas*

1. The cost of water supply for the purpose of water tariff fixation shall consider the cost of production and supply of water,

comprising the capital cost of the system and operation and maintenance costs; the cost of resource depletion; and environmental degradation costs.

2. Water used for economic production functions, such as industrial and commercial uses, should be charged the highest, while water for human needs (drinking and domestic needs) should be charged the lowest.
3. Water required for lifeline consumption in the domestic sector should be charged the least, with higher charges for higher levels of use.
4. The water tariff for bulk users, such as multi storeyed apartments and other housing stocks, shall consider the number of families served while working out the differential water rates for consumption in excess of lifeline needs.

Article 26: *Norms for fixing water tariffs in rural areas*

1. The cost calculations for the purpose of water tariff fixation shall consider only the operation and maintenance costs if the cost of production and supply of water is very low for the scheme.
2. If the cost of production and supply of water is high for the scheme, then the cost calculations for water tariff fixing should include the capital cost of the system and the O&M costs.
3. If individual household level metering is practised, lower unit charges shall be recovered for the minimum level of consumption to meet the basic survival needs from all households with metered connections, and higher unit rates for higher levels of consumption.

Article 27: *Rules for metering of water supplies in rural and urban areas*

1. Metering of water supply is essential for implementing a sound water tariff policy based on volumetric consumption of water.

2. Metering of water supply involves measuring the volume of water supplied from the source as well as the volume of water used by consumers for fair pricing and transparency in the operation of the Public Water Utilities.
   The Public Water Utility shall install water meters at every entry point of water mains leading to a building, and water meters are considered components of the indoor water supply infrastructure of a building.
3. Water meters, regardless of their place of installation, shall be considered the property of the client.
4. Consumers are liable to pay for the cost of meters, their installation and the cost of replacement of meters from time to time, and they must also report malfunctioning meters to the utility.
   The rules relating to the installation, operation and maintenance of water meters, replacement of meters and sanctions for violations of the rules shall be prepared by the concerned Public Water Utility.
5. It is the duty of the Public Water Utility to ensure that the installed meters meet the quality standards as specified by the rules.
6. Consumers shall purchase water meters of any brand specified by the department and shall produce a certificate of testing from an authorized laboratory at the time of release of connection.
7. For water connections of 50 mm and above for domestic purposes and for water connections for commercial and industrial purposes, automatic digital meters or meters as per the standard decided by the Public Water Utilities shall be installed.
8. The work related to metering, billing and collection of water charges should preferably be outsourced.
9. Violation of the rules pertaining to the operation and maintenance of water meters and water counters shall be punishable under the Act.

3.8. Chapter 8: Water Supply Technologies

Article 28: *Criteria for selection of water supply technologies in urban areas and major projects (with costs exceeding Rs. 25 crore) of Rajasthan*

1. The selection criteria for Water supply technology refer to the broad range of considerations involved in the selection of a water supply technology from the available options. These criteria can change from situation to situation, from locality to locality, and from time to time, as well as according to the policies of the government.
2. The criteria for selecting water supply technologies shall include physical, economic, technical, financial, institutional and environmental factors – such as water availability in the hydrological system, water quality, source reliability, seasonal changes and source protection.
3. Given the scarcity of fresh water resources in Rajasthan, particularly in large parts of the state with high variability in rainfall, stream flows and groundwater recharge, the 'Utilities' should explore all possible technologies to ensure sustainable supplies.
4. Environmental conditions and the physical and economic sustainability of water supply services will be the three most important considerations driving the selection of water supply technology.
5. The life cycle cost of the system shall be considered in evaluating its economic sustainability.
6. Physical, economic and environmental factors shall be non-negotiable, whereas technical, financial and institutional factors shall be negotiable.
7. In situations where dependable and reliable water sources cannot be found from surface catchments or underground formations, options such as rainwater harvesting, inter-basin water transfers and treatment of domestic wastewater should be explored.

Article 29: *Criteria for selection of water supply technologies in rural areas of Rajasthan*

1. In rural areas, in addition to physical, economic, financial, institutional and environmental factors, the local availability of materials for construction and technical skills for maintenance and repair of the system should also be considered when selecting a water supply technology.

Article 30: *Rules and incentives pertaining to the construction of rooftop rain water harvesting structures for water collection and recharge in rural and urban areas*

1. Rooftop rainwater harvesting will be mandatory for residences, commercial buildings, institutional buildings, and industrial buildings in both rural and urban areas, for either storage or groundwater recharge, if:
   (a) The minimum per capita roof area is above 5 sq. m in the case of apartment buildings, and the total roof area is equal to or above 50 sq. m for individual dwellings;
   (b) The total roof area of the building is 100 sq. m or more in the case of commercial buildings and industrial units.
2. In rural areas and slums in urban areas, rooftop water harvesting will not be mandatory for Kachchha houses and BPL households.
3. Construction of rooftop water harvesting structures shall be undertaken for groundwater recharge only if the water in the underground formation is of good quality and the aquifers have storage potential. Otherwise, the water from the roof catchment shall be stored in storage tanks for direct use.
4. For new buildings, the issuance of a license shall be contingent upon the owners constructing rooftop water harvesting systems as per the approved designs of the municipality or the Public Water Utility, if the buildings meet the criteria with respect to roof area.
5. For already existing buildings that meet the above criteria for roof area, the ownership deeds shall be cancelled, and the water,

electricity and sewer connections will be removed if the owners fail to install the RWH system on the premises within six months from the date of notification.

6. Financial incentives in the form of rebates on water tariffs shall be provided to households with rooftop rainwater harvesting systems that receive water supply from the utility, provided their connections are metered and they pay for water on a volumetric basis.

3.9. Chapter 9: Location, Design and Construction of Water Supply Systems

Article 31: *Rules and guidelines for the location of water supply systems*

1. For the location of water supply systems, the existing guidelines in the latest CPHEEO manual and circulars issued by the Public Water Utilities from time to time shall be followed.
2. In the case of surface water-based schemes (such as those based on reservoirs, lakes, ponds, rivers/streams and canals), the location of the point of water abstraction shall be such that water of sufficient quality in terms of physical, chemical, microbial and radiological properties can be obtained with minimum treatment.
3. In the case of groundwater-based water supply systems, the location of the pumping well or borehole of the hand pump should be at a safe distance from landfill sites, land-based wastewater treatment plants, lakes/ponds, natural drainage systems receiving sewage and industrial effluents, and septic tanks.
4. If needed, the above-mentioned 'safe distance' shall be ascertained by the Utility through scientific studies.
5. For special structures and innovative technologies that are neither covered in the CPHEEO manual nor in the BIS codes, such cases shall be referred to the Technical Committee of RWSSMB.

Article 32: *Rules and guidelines for design of water supply systems*

1. For the design of water supply systems, the existing guidelines in the latest CPHEEO manual and circulars issued by the Public Water Utilities from time to time shall be followed.
2. In the case of surface water-based systems, the system should be designed for 95 per cent dependability based on past stream flow records, after making allowance for allocations already made for competing sectors of water use.
3. The design should also consider all future developments of surface water and watersheds in the upper catchment by various agencies, which were approved at the time of planning.
4. The design of such schemes should consider all interventions for the protection of the catchment to ensure a water yield of sufficient quantity and quality.
5. The design of groundwater-based schemes should be based on the safe yield that can be obtained in the driest season.
6. For surface water-based schemes planned for a population above 4000 people, proper hydrological studies, including those using simulation models, shall be undertaken to ascertain catchment yields.
7. For groundwater-based schemes planned for a population above 4000 people, proper geo-hydrological studies, including pumping tests and modelling studies, shall be undertaken to ascertain the specific capacity and safe yield of the aquifers.
8. For special structures and innovative technologies that are neither covered in the CPHEEO manual nor in the BIS codes, such cases shall be referred to the Technical Committee of RWSSMB.

Article 33: *Rules and guidelines for construction of water supply systems*

1. For the construction of water supply systems, the existing guidelines in the latest CPHEEO manual and relevant circulars issued by the Public Water Utilities from time to time shall be followed.

2. For the procurement of materials for the construction of water supply systems, the existing guidelines in the latest CPHEEO manual and relevant circulars issued by the Public Water Utilities from time to time shall be followed.
3. The construction of water supply systems should conform to the guidelines and directions issued by the Public Water Utility from time to time.
4. The distance between drinking water supply pipelines and sewer lines shall be a minimum of one metre horizontally and vertically.

3.10. Chapter 10: Operation and Maintenance of Water Supply Systems

Article 34: *Rules and guidelines for operation and maintenance*

1. For the operation and maintenance of water supply systems, the latest CPHEEO manual and relevant circulars issued by the Public Water Utility from time to time shall be followed.
2. The lifespan of various components of water supply systems shall be defined by the Public Water Utility and they must be replaced once their lifespan expires.
3. The operating efficiency of the different components shall also be defined by the Public Water Utility.
4. The minimum qualifications and experience of the technicians engaged in operation and maintenance of the water supply systems shall be prescribed by the Public Water Utility for each type of water supply system.
5. Matters not covered under the CPHEEO manual shall be referred to the Technical Committee of RWSSMB, and the decision of the committee shall be final and binding.

Article 35: *Right of way for land acquisition for construction and maintenance of different components of water supply systems*

1. An adequate amount of land, whether private or public, shall be made available under this Act for the construction of water sources, water distribution systems, treatment plants, intake structures, ground and overhead reservoirs, and any other component of the water supply system, as per the approved plan, design and drawing.
2. An adequate amount of land shall be made available under this Act for carrying out the operation and maintenance of water supply systems for an adequate period.
3. In the case of physical damage to any component of the water supply system by any person, group of persons or agency, they shall be liable to pay compensation to the department. If the person, group of persons or agency fails to pay the compensation as decided by the Public Water Utility, the Utility may disconnect their service connection and take legal action.

3.11. Chapter 11: Recycling and Reuse of Wastewater

Article 36: *Rules and regulations pertaining to the recycling of wastewater*

1. 'Recycling' of treated wastewater shall be done in situations where the fresh water resources are 'absolutely scarce' in an area, and water from the same area is used for industrial processes that generate hazardous liquid waste in large volumes, and disposal of the same can potentially cause serious environmental hazards.
2. The wastewater from such processes should be subjected to tertiary treatment for the removal of hazardous chemicals (heavy metals) and microbial contaminants before recycling.

3. The disposal of untreated or partially treated wastewater generated from manufacturing processes (including those with secondary treatment) in such areas shall be a punishable offence under this Act.
4. The use of treated wastewater from industries and commercial units (hotels, restaurants and utility buildings) that conforms to the intake water quality in terms of physical and chemical quality and microbial content can also be considered 'recycling'.

Article 37: *Rules and regulations pertaining to the reuse of wastewater*

1. Reuse of wastewater concerns the use of wastewater that is subject to a minimum of secondary treatment.
2. Any wastewater that is of inferior quality cannot be considered for 'reuse'.
3. Reuse of treated wastewater, including landscape irrigation, shall be allowed only in situations where the activity doesn't cause environmental problems, even if the quality of the wastewater conforms to effluent quality guidelines for human health and crop protection stipulated by WHO (1989).
4. The supply of treated wastewater for reuse by the Public Water Utility shall use a separate pipeline system from the one used for water supply.

Article 38: *Incentive and disincentives for promoting recycling and reuse of treated wastewater*

1. Special incentives in the form of tax exemptions can be offered to industries that are adopting water recycling measures in their industrial processes.
2. The amount of a tax incentive can be determined by the amount of wastewater treated and recycled, based on certification from the State Pollution Control Board.

3. The State Pollution Control Board will be the competent authority to enforce compliance with norms on wastewater recycling and reuse and issue certification, and it shall have the power to fix penalties for violations of various provisions of the Act under Articles 36 and 37.
4. Recycling and reuse of domestic wastewater by households within their dwelling premises shall be incentivized through rebates in the water tariff.

3.12. Chapter 12: Water Supply Surveillance

Article 39: *'Targeting' in water supply surveillance*

1. 'Water supply surveillance' concerns the monitoring of the outcomes of water-related services provided by the Public Water Utility mainly in urban areas.
2. 'Water supply surveillance' should be undertaken regularly in areas that are high risk and highly vulnerable to 'water-related health hazards', to prevent public health issues associated with water contamination and poor sanitation.
3. Cities and towns in regions experiencing frequent droughts and acute water scarcity, where the degree of access to water supply from formal sources is low and access to sanitation is poor should receive high priority in 'water supply surveillance'.
4. Within these cities and towns, the surveillance activity shall concentrate on areas that are densely populated, especially slums, tail-end points of the distribution system, low-lying areas and areas at high elevations that are prone to flooding.
5. In other urban areas, surveillance activity shall also be undertaken if there are pockets in these areas that have situations similar to those mentioned above.

Article 40: *Protocols for 'surveillance of water supply' for prevention of public health hazards and risks*

1. Surveillance activity in any area should be based on the preparation of a water safety plan and its approval by the competent agency, and should report on the following:
   (a) Progress in the water supply and sanitation situation: service levels; water pressure; water quality parameters (fluoride, chlorides, pH, hardness, salinity; microbial quality).
   (b) Incidence of leakage in water distribution pipes.
   (c) Breakage in sewerage lines and overflowing of sewers.
   (d) Open defecation.
   (e) Incidence of water contamination in the pipelines with chemical and microbial contaminants.
   (f) Incidence of outbreaks of water-borne and vector-borne diseases and the number of people affected.
2. The frequency of surveillance activity (surveys) should be decided by the degree of vulnerability of the urban area and localities within the urban area under surveillance.

Article 41: *Agencies responsible for water supply surveillance*

1. 'Water supply surveillance' should be undertaken only by state-level institutions, outside the government that are autonomous and have the required scientific knowledge and institutional capability to undertake such surveys.
2. The surveillance activity should be financed by the Ministry of Water Supply of the state government.
3. The surveillance agency should not have any 'conflict of interest' with the water service institution that serves the area under surveillance.
4. The surveillance activity shall be monitored by an independent body of water supply and public health experts, chaired by the Principal Secretary, Water Supply, for corrective measures.

5. In rural areas, community-based water supply surveillance shall be undertaken using 'field test kits', as prescribed under the latest NRDWP guidelines.

3.13. Chapter 13: Functioning of 'Water Service Providers'

Article 42: *Rules and guidelines on the functioning of water service providers for drinking water purposes*

1. All 'water service providers' shall be registered with the *Food Commissioner* of the State of Rajasthan.
2. The water service provider should possess a Food Business Operator (FBO) license under the Food Safety and Standards (Licensing and Registration of Food Business) Rules, 2011.
3. In the case of water tankers, the FBO license number should be displayed on the tankers. Establishments using multiple vehicles to supply water are required to obtain a special license that contains the registration numbers of all vehicles.
4. All drinking water sources owned by the water service provider (for both tanker supplies and packaged drinking water) must have the FBO license.
5. For tanker water supply, the vehicles used for supplying drinking water must carry signage in Hindi.
6. The interiors of the water tanks must be coated with certified materials in order to prevent contamination of the water.
7. Every six months, the quality of the water from the source should be tested at government laboratories or those accredited by the National Accreditation Board for Testing and Calibration Laboratories.
8. In the case of packaged drinking water, only 'certified materials' conforming to relevant BIS codes shall be used for packaging. The quality of the water (physical, chemical and microbial) shall be displayed on the pouch, bottle, or can used.

Article 43: *Registration of water service providers*

1. The water service provider who wants to get registered with the Food Commissioner shall furnish details such as the assets owned; the sources from which they acquire water; the types of water treatment and packing facilities (as the case may be) owned and their capacities; the types of water conveyance and transportation systems; the amount of skilled and unskilled manpower available to them; and the nature of services.
2. The tankers engaged by the Public Water Utility should install GPS systems at their own expense to enable the Utility to monitor and track their movements.

Article 44: *Rules and guidelines on the functioning of water service providers for 'uses other than human consumption'*

1. Clauses 1, 2, 3 and 7 under Article 42 will apply in the case of water being supplied by private water providers for purposes other than human consumption.
2. For those vehicles transporting water for purposes other than human consumption, the signage must state 'Not for Drinking Purpose' on the tanks.

Article 45: *Sanctions against water service providers for violation of the rules*

1. Violation of the rules pertaining to the functioning of water service providers shall attract sanctions.
2. The sanctions for non-compliance shall include penalties up to cancellation of licence and prosecution under various provisions of the Food Safety Act 2006 and the Food Safety and Standards Rules 2011, Government of India.

3.14. *Chapter 14: Illegal Water Tapping, Drinking Water Monitoring, Protection of Drinking Water Sources, and Contamination and Pollution of 'Drinking Water Sources'*

Article 46: *Rules and regulations for the protection of drinking water sources*

1. 'Drinking water sources', as referred to in the Act, shall be designated by a competent authority in the Public Water Utility and shall include only the raw water source, which will include springs, hand pumps, tube wells, bore wells, open wells, reservoirs (both man-made and natural), ponds, dams, canals and rivers.
2. 'Drinking water monitoring' covers water quality and treatment performance, including source water, process performance, treated water and water in the distribution system. Protection of the source includes protection from illegal water tapping and protection of the water from pollution and contamination.
3. 'Drinking water monitoring' and protection of drinking water sources in an area shall be the sole responsibility of the Public Water Utility, which owns the source.
4. 'Drinking water monitoring' is to be undertaken to determine whether the water treatment system and water distribution system are performing at the appropriate level as per their design.
5. 'Drinking water monitoring' requires that water samples be collected regularly from the source and tested in government-accredited laboratories to see whether the quality of the water conforms to the accepted standards, as per the rules framed for the purpose.
6. 'Protection of drinking water sources' requires observing safety protocols as per the rules framed for the purpose by the Utility.
7. Contamination and pollution of 'drinking water sources' and illegal water tapping shall be treated as punishable offences. On detection of foul play by any individual identified or otherwise, the Public Water Utility shall file a complaint with the police department.

8. Vigilance teams of the departmental officials shall be constituted with the support of the police for checking water theft from 'drinking water sources', and illegal water connections.

Article 47: *Regulation of groundwater development for protection of drinking water sources for both quantity and quality*

1. This concerns regulation of groundwater development by private individuals/agencies or by public agencies for any purpose, including drinking water supplies for private purposes, aimed at protecting drinking water sources managed by Public Water Utilities.
2. Under this Act, the Public Water Utility will have the power to restrict or stop the development of groundwater resources, if the situation warrants, to protect surface as well as groundwater-based drinking water sources. Such situations are those where new wells are developed in the influence area of either surface water or groundwater-based drinking water sources after the source was created by a Public Water Utility.
3. The distance of the new well from the drinking water source shall be a minimum of 200 m and a maximum of 500 m for alluvial areas and 500 m to 1,000 m in the case of hard rock areas. The same shall be notified in the area where a new drinking water source is created for public awareness.
4. The State Electricity Utility shall seek a No Objection Certificate from the Public Water Utility before offering new electricity connections for groundwater wells in an area to ensure that the new source is not in violation of the regulation.
5. Information shall be furnished by the applicant for electricity connections for operating a well, on the capacity of the well (depth, diameter and pump capacity) and its distance from the nearest existing drinking water source to both the Public Water Utility and the State Power Utility.

6. If an act of violation of the said regulation is established, wherein a private person, a private agency or a public agency is found abstracting water from a well using any mechanism, a competent authority from the Public Water Utility shall seal the well, confiscate the water abstraction devices, including the pumping machinery, and impose stringent penalties.
7. In case the owner/owners of the wells do not cooperate with the officials, the PWU shall seek the help of the police to carry out raids.

Article 48: *Criteria for fixing fines by the Public Water Utility for non-payment of water charges*

1. The fine for non-payment of water charges by consumers in situations where they are otherwise not liable for disconnection shall be proportional to the payment overdue to the Utility, the amount of delay in payment, the interest rate that the Utility likes to charge for the period of overdue, and a suitable penalty.
2. The Utility shall charge an interest rate equal to two times the bank interest rate.

Article 49: *Criteria and norms for fixing penalties for illegal water tapping*

1. As per this Act, both theft of water from 'drinking water sources' and illegal connection to a water supply system managed by a Public Water Utility fall within the ambit of 'illegal water tapping'.
2. On detection of illegal water tapping, the Public Water Utility may impose a penalty on the illegal water user, to be fixed by a competent official from the Utility, based on the revenue losses to the Utility and estimates of the volume of water illegally tapped, arrived at using the information at its disposal.
3. Failure on the part of the illegal water user to pay the penalty within the stipulated time period as notified will enable the Utility to seek legal remedy in a court of law.

Article 50: *Legal actions for non-compliance with drinking water monitoring and source protection and for contamination and pollution of drinking water sources*

1. Non-compliance with the measures for monitoring and protection of water sources by any Public Water Utility shall invite legal action under the Food Safety Law 2006 and the Food Safety and Standards Rules (2011) therein.
2. The person indulging in illegal tapping of water from a source or acts resulting in contamination and pollution of drinking water sources shall be liable for prosecution under the Food Safety Act 2006 and various provisions of the CrPC.
3. In the case of manhandling of staff of the Utility by any individual/group of persons and/or the creation of 'law and order' problems arising out of unforeseen situations or technical failure of the water supply system, the Utility shall ask for police protection for its concerned staff, and it will be incumbent upon the police department to provide such protection on a priority basis.

Article 51: *Agencies empowered to impose penalties on defaulters*

1. Under this Act, the Public Water Utility shall be empowered to take action against defaulters of water tariff payments and those engaged in illegal water tapping from the distribution system in the form of penalties.
2. In a situation where the defaulter is not cooperating with the Utility, the Utility would have the option of seeking legal remedy in a court of law.

3.15. Chapter 15: Institutional Capacity Building for Water Supply

Article 52: *Criteria for evaluating the performance of Public Water Utilities*

1. The evaluation of the performance of Public Water Utilities shall use broad criteria and therefore be: technical, physical, economic, institutional, managerial and environmental.
2. The technical criteria shall include the condition of the drinking water source and intake structures and the engineering components for raw water treatment, water supply, sewage collection and wastewater treatment.
3. The physical criteria for performance assessment shall consider the average per capita water supply, coverage of water supply, quality of water, sewerage system coverage and extent of treatment of wastewater.
4. The economic criteria for performance assessment shall consider the cost per unit of water produced and supplied. The lower the cost, the higher the economic performance, subsidies apart.
5. The criteria for managerial performance of the Utility are the number of water and sewerage connections per staff and the amount of water supplied per staff.
6. The criteria for environmental performance of the Utility are: the amount of water depleted and degraded against every unit of freshwater supplied by the utility at the end-user level.
7. The institutional performance of the utility shall be assessed in relation to equity in access to water supply, in terms of both physical access and affordability of water, and the degree of social inclusion.

Article 53: *Water auditing and non-revenue water*

1. The purpose of water audits shall be to continuously improve the performance of public water supply systems and utility performance.

2. Water auditing shall be made mandatory for big consumers with more than 50 mm connections for commercial and industrial use.
3. The Public Water Utility shall engage independent agencies to undertake water auditing programmes for all towns.

Article 54: *Norms on water quality monitoring and sampling and testing of water*

1. A Water Safety Plan shall be prepared for every drinking water supply system in rural and urban areas.
2. Water quality monitoring shall be undertaken by the Public Water Utility or the water supply institution wherever it exists for the drinking water sources, the treated water, and the water supplied at the end of the pipe against all water quality parameters, viz., physical, chemical, microbial and radiological, to prevent contamination of the sources and to ensure compliance with the drinking water quality standards in the treated water and water supplied through the distribution system.
3. Water quality testing laboratories shall be set up in every district and sub-district, with facilities for testing water samples for all physical, chemical, and microbial quality standards to which the drinking water quality should conform.
4. Water quality management (WQM) shall conform to the uniform protocols for WQM, including the water quality parameters to be tested, the frequency of monitoring, drinking water sampling and water testing, issued by the Ministry of Drinking Water and Sanitation, Govt. of India, from time to time.
5. However, WQM of 'drinking water sources' and water supplied at the end of the pipe for microbial quality shall be undertaken once a month in the case of surface sources such as water received in reservoirs from the natural catchment or water in streams and canals.
6. For groundwater-based sources, the frequency of monitoring for chemical and microbial quality of water shall be once every three months, including pre- and post-monsoon.

7. Testing of the quality of drinking water for heavy metals and pesticides shall be undertaken if the need arises.

Article 55: *Framework for institutional reforms in the drinking water supply sector*

1. Institutional reforms in the drinking water sector shall be undertaken to build the capacities of the various institutions engaged in 'water resources management', including water quality management and drinking water supply services, to manage the supply of drinking water to consumers sustainably, efficiently and equitably.
2. Institutional restructuring and redefining the nature of interactions between institutions engaged in public water supply will be part of the institutional reform process.
3. Institutional reforms shall lead to greater transparency and accountability and better incentives among the various agencies in the institutional hierarchy to enhance their performance in their respective line functions. These agencies include Public Water Utilities, water supply institutions, institutions for water resources management created at the catchment or river basin-scale, and the State Pollution Control Board.
4. The key institutional reform measures that shall be undertaken in the short term are:
   (a) Metering of water supplies at the level of bulk suppliers, the water utility, and at the consumer level.
   (b) Volumetric pricing of water.
5. The key institutional reform measures in the medium and long term are:
   (a) Introduction of a pollution tax for disposal of untreated urban sewage and industrial effluents.
   (b) Coordination of the activities of various water supply service management agencies, viz., the Water Resources Department, Public Water Utility, and the Industrial Development Corporation, by the river basin authorities.

(c) Separation of line agencies for water quality monitoring in natural water bodies from those responsible for enforcement of pollution control norms to check the performance of agencies responsible for WQM and the health of natural water systems.

Article 56: *Guidelines for strengthening organizational setup*

1. The roles and responsibilities of each post in the Public Water Utilities shall be defined, along with expected outputs and deliverables.
2. Staffing of all Public Water Utilities shall be reviewed every year, and recruitments and promotions shall be undertaken annually commensurate with the job requirements for proper cadre management. This is essential for delivering proper service to people in departments that handle emergency services.
3. The Public Water Utilities may seek an external review of their administrative and financial performance undertaken by competent agencies and obtain specific suggestions for improving their performance.
4. The Public Water Utilities shall involve NGOs and CBOs in select activities (as specified under Articles 59 and 61) within the ambit of their work to improve efficacy, and adequately compensate them for the services rendered.
5. A separate institution shall be created by the PHED for promoting local institutions to ensure community participation in drinking water management, and the existing Water and Sanitation Support Organization may be upgraded on the lines of WASMO Gujarat, as prescribed in NRDWP guidelines.

Article 57: *Guidelines on human resource development of drinking water administration*

1. Human resources development involves providing technical assistance to develop the capacity of local institutions such as

Municipalities/Public Water Utilities/PRIs/VWSCs for various administrative, financial and managerial activities to perform water management functions effectively.

2. Human resource development in drinking water administration should reflect the emerging needs of the sector for improved performance.
3. HRD programmes should enable technical and managerial staff at all levels in the Public Water Utilities and other water institutions to effectively perform their duties.
4. Human resource development shall be a regular and ongoing activity in the Public Water Utilities to equip staff to meet new scientific, technical, managerial, financial, legal and institutional challenges in the drinking water supply sector.
5. HRD programmes shall cover all levels – operators, technicians, engineering staff, techno-managerial staff and senior management staff – and shall include in-house training as well as external training.
6. HRD programmes shall cater to the members of water supply institutions based on identified training and capacity-building needs.
7. An annual assessment of the training needs of Public Water Utility staff shall be conducted at all levels.
8. The Public Water Utility shall engage established government and private research and training institutions in the water supply sector as 'consultants' for both 'training need assessment' and conducting training programmes.

Article 58: *Types of HRD programmes*

1. The types of HRD programmes for the staff of Public Water Utility and water supply institutions shall be decided based on the nature of duties currently performed or likely to be performed in the future by them.

2. Training will be mandatory once every six months for all engineers and technical staff. The training will cover technical, economic, social, financial and managerial issues.
3. The trainings will be conducted by the Water and Sanitation Support Organization with the assistance of established key resource centres.
4. The training programmes for Public Water Utilities should broadly cover the following areas:
   (a) Operation and maintenance of water supply schemes
   (b) Design of water supply systems using advanced technologies such as desalination, main system management, SCADA (Supervisory Control and Data Acquisition); water auditing; application of leakage detection and reduction technologies
   (c) Efficient water use technologies in dwellings and commercial establishments to conserve water
   (d) Water pricing and management of non-revenue water
   (e) Technical and economic aspects of wastewater treatment technologies
   (f) Water quality monitoring and water supply surveillance
   (g) Recycling and reuse of wastewater
   (h) Green infrastructure and stormwater management systems
   (i) Rainwater harvesting and artificial recharge systems

Article 59: *Guidelines on the involvement of NGOs in managing drinking water supply schemes*

1. NGOs and other civil society organizations shall be involved by the Public Health Engineering Department or the WSSO in undertaking activities at the village level, such as capacity building, social mobilization, and IEC (Information, Education and Communication) to facilitate the formation of VWSCs and Pani Panchayats and impart skills to the members of VWSCs and Pani Panchayats, with necessary financial support provided by the line agencies.

2. The eligibility criteria for the selection of these organizations (NGOs and other civil society organizations) for the assignments shall be decided by the Public Water Utilities.
3. The ToR (Terms of Reference) and the scope of work for these organizations for undertaking IEC activities and training programmes shall also be determined by the concerned line agency.

Article 60: *Rules and regulations on private sector participation in drinking water supply management*

1. The goal of private sector participation in drinking water supply management is to strengthen the organizational capacity of public water utilities.
2. Geographically and administratively, private sector participation in drinking water supply management shall be considered in both urban and rural areas.
3. Private sector participation shall be sought only in situations where the Public Water Utilities lack adequate technical, managerial and administrative capabilities.
4. The private sector's participation shall be in the following areas: design, construction and operation of water treatment and water supply systems; wastewater treatment plants; design of stormwater drainage and sewerage systems; design, construction and operation of desalination systems; leakage detection and prevention works; and any other area involving sophisticated and innovative technology.
5. Private sector participation could take the form of service contracts, operation and maintenance contracts, lease agreements, concession agreements, BOT (build-operate-transfer) agreements, and joint ventures, depending on the scope and nature of the work.

Article 61: *Rules and regulations for Village Water and Sanitation Committees and Pani Panchayats for drinking water management in rural and urban areas*

1. Village Water and Sanitation Committees will have legal powers to operate and manage individual village-based water supply schemes.
2. The operation and management of individual village-based water supply schemes will involve regular operations and routine maintenance works, ensuring equitable water distribution, setting water tariffs for financial viability; coordinating major repairs with the local wing of the water supply utility, and collecting water charges.
3. For rural regional water supply schemes serving multiple villages, the Public Water Utility will handle regular operations and routine management up to the village entry point, while VWSCs at the village level will manage within individual villages.
4. In regional water supply schemes, responsibilities of village-level VWSCs include operating and maintaining water distribution within the village, ensuring equitable distribution among households, coordinating major repairs with the local water supply utility, and collecting water charges.
5. The Village Water Supply Committee may hire a support organization for O&M of water supply schemes within the village, if necessary.

Article 62: *Rules and regulations on the functioning of PRIs for drinking water management in rural areas*

1. The Gram Panchayat will facilitate community involvement in planning new water supply schemes in the village and will organize the Gram Sabha meetings to introduce NGOs designated for community mobiliziation.

2. The Gram Panchyat will govern individual village schemes, including issuing household water connections, setting connection charges, fixing water tariffs, determining water supply duration, and imposing penalties for non-payment of water charges.
3. For schemes spanning multiple villages and crossing Gram Panchayat jurisdictions, higher-level local administrations (block Panchayat or district Panchayat) will oversee governance.
4. The fifth welfare committee of the Gram Panchayat will be integrated into the Village Water Supply Committee, as per Principal Secretary Order No. F-24/SWSM/10–11/854–89 dated 31 May 2010.

3.16. Chapter 16: Guidelines and Criteria for Convergence of Government Programmes in Rural Areas

The articles under this chapter focus on setting guidelines and criteria for allocating funds from various government programmes aimed at public works in rural areas. These programmes include watershed development, rural employment guarantee schemes, water conservation/artificial recharge schemes and other rural development initiatives. The goal is to promote sustainable water supplies and environmental sanitation in rural areas facing issues such as drinking water scarcity and inadequate liquid waste disposal.

Article 63: *Guidelines for convergence of government programmes*

1. In selecting activities under various government programmes involving public works in villages with water scarcity issues, priority should be given to projects that enhance water supply in terms of quantity, quality and accessibility when allocating funds.
2. Activities requiring significant capital investment, including materials such as cement, bricks, RCC, sand, gravel, concrete, stainless-steel pipes, pumping machinery and RO membranes, may also be considered under these programmes.

3. When selecting activities under government programmes for villages facing sanitation challenges, priority should be given to initiatives that improve environmental sanitation when distributing funds.
4. The activities selected in villages facing poor sanitation problems shall also include those that involve significant costs for capital works and the use of materials such as cement, sand, bricks, concrete, gravel, cement pipes and pumping machinery.

Article 64: *Criteria for convergence of government programmes*

1. The villages where less than 50 per cent of the households have access to an improved source of water supply shall be considered 'villages facing acute drinking water shortages'. Similarly, villages where less than 50 per cent of the households have access to improved sanitation shall be considered 'villages facing problems of poor sanitation'.
   The activities chosen under these government programmes in the villages facing drinking water scarcity shall directly contribute to improving public water supply for the households, in terms of water quantity water quality, physical access to water supply sources, or all of these aspects.
2. The activities chosen under the government programmes in villages facing problems of poor sanitation shall directly contribute to enhancing access to improved sanitation for all.
3. A minimum of 25 per cent and a maximum of 50 per cent of the total funds available from all such government schemes for the villages shall be diverted to water supply and sanitation programmes on a priority basis.
4. None of the funds allocated shall be used for creating physical assets that benefit individual households, unless the government programme or scheme prescribes the use of funds for creating such assets.

5. Prohibited works include the creation of private toilets, private water supply connections from public systems, and individual household-level water treatment systems and soak pits.

3.17. Chapter 17: Agency for Bulk Water Supply

Article 65: *Creation of a bulk water supply agency*

1. Under this Act, a bulk water supply agency shall be promoted as an institution for the allocation of water from a single or multiple surface water sources among several competing users.
2. The state government shall constitute a bulk water supply agency. This agency will coordinate among different competing users for the allocation of water from the surface water source.
3. It will be an autonomous institution, chaired by a bureaucrat in the rank of Principal Secretary/Secretary to the state of Rajasthan. The management board of the agency shall consist of water resource experts and other related discipline experts, and it shall be constituted by the Government of Rajasthan.
4. A bulk water supply agency will be allowed to function on a commercial basis, and it shall have the power to handle only water supplies from large water sources with competing claims.

Article 66: *Functions of a bulk water supply agency*

1. The main function of the bulk water supply agency will be to manage the inter-district allocation of water from a source among different 'claimants' of the water from the said source and to fix the price of water.
2. The claimants of water from the source will only hold 'water use rights', and not absolute ownership rights over the water from the source, based on the average water use of the claimant for the previous five years from the said source/sources.

3. It will not be obligatory for the bulk water supply agency to honour the 'claims' of any right holder if the water available in the source/sources in a particular year is less than the lowest during the previous five years.
4. The bulk water supply agency will allocate water from the source/ sources to different 'claimants' based on the 'demand', and the available quantum of water during every hydrological year. The allocation shall under no circumstances exceed the maximum water use by the 'claimant' in any of the previous five years.
5. The bulk water supply agency will also decide the price at which water will be sold to different users based on the average value of water.

Article 67: *Duties and responsibilities of the water right holders*

1. It is the duty and responsibility of the water right holders to make claims on water based on actual demands.
2. In case of need, the claimant shall cooperate fully with the bulk water supply agency to verify its demands, which are made in writing to claimants and a competent official of the agency.

3.18. Chapter 18: Water, Sanitation and Hygiene

Article 68: *Promotion of improved sanitation in rural areas*

1. Investments in water supply by the Public Water Utilities in both rural and urban areas should promote improved sanitation and water supply systems, with supply norms determined accordingly.
2. In rural areas, individual household-level water supply connections shall be encouraged through financial incentives in the form of subsidies for water connection charges if the households have individually owned toilets within the dwelling premises.
3. The amounts of subsidies for the households shall be decided by the Public Water Utility, and the same shall be paid by the Utility

to the rural water institution, which is mandated to collect connection charges from the households.

4. The rural water institution, in turn, shall offer the subsidy to the household that applies for a new water supply connection.

Article 69: *Promotion of improved sanitation in urban areas*

1. In urban areas, financial incentives shall be provided by the Public Water Utility to the housing stock (apartment complexes, bungalows), commercial establishments and utility buildings in the form of discounts in the water tariff if they have built their own domestic wastewater treatment systems and reuse systems with zero discharge into the sewerage system.
2. Quantitative criteria shall be used for fixing the discount in the water tariff based on the quantum of wastewater treated and reused.
3. The exact discount in tariff to be offered to the water users will be decided by the Utility.

Article 70: *Hygiene promotion*

1. Once a new water supply scheme is made operational in an area, either rural or urban, the Public Water Utilities should organize awareness activities aimed at hygiene promotion to maximize the health and welfare impacts of improved water supply.
2. The hygiene promotion activities can be undertaken either by the staff of WSSO, or by the 'water supply institutions' or NGOs engaged by WSSO for the purpose.
3. Special funds shall be allocated to schools to conduct education and awareness programmes on hygiene promotion by the Public Water Utility.

4. Conclusions

The draft Drinking Water Services Act presented in this chapter for the state of Rajasthan, presented in eighteen chapters and seventy articles, consists of the following: (1) a listing of terms used and their definitions; (2) the general provisions; (3) guidelines, criteria, norms pertaining to water supply service levels, including quantity quality and frequency of water supply; treatment of raw water for public water supply; issuing of water supply connections; water pricing; selection of water supply technologies; location, design and construction of water supply systems; operation and maintenance of water supply schemes; treatment and disposal of wastewater; recycling and reuse of wastewater; measures for prevention of Illegal water tapping, drinking water monitoring, protection of drinking water sources and contamination & pollution of 'drinking water sources'; institutional capacity building in the drinking water sector for improved performance; functioning of water service providers; water supply surveillance; and hygiene promotion; and (4) functions, duties and responsibilities of the bulk water supply agency.

The definition of various terms used in the draft Act, presented in the first chapter, is to prevent any kind of ambiguity that can occur while the articles are interpreted, thereby avoiding misinterpretation of the 'Act'. Each article under different chapters carries a 'title', which indicates what that particular article deals with.

CHAPTER 5

How to Make Water-Related Acts Work Better in India

1. Introduction

Researchers have argued for the need to create new legal instruments to improve equity and sustainability in water use in India at the state level, following the legal and policy frameworks set at the national level (see, for instance, Cullet, 2007, 2016). However, there is no systematic scholarly attempt to critically examine the reasons for the ineffectiveness and failure of existing or past legislations related to water, with the exception of Phansalkar and Kher (2006), who examined the erstwhile Groundwater Act (1993) of Maharashtra.

In this chapter, we look into the factors that contribute to the ineffectiveness or complete failure of water-related Acts in India in a way that helps frame more meaningful and enforceable legislation. Since water is a state subject in India (except for the inter-state rivers, which fall on the concurrent list), we examine the water-related Acts framed by the state governments. The chapter does not provide a comprehensive review of the performance of specific water-related Acts. Instead, it identifies certain unique features that are characteristic of Indian water resources or services Acts and that become the recipe for their failure, ineffectiveness or redundancy. The basic highlight of the chapter is the need for having rigour in understanding the complex problems and issues, as such rigour is critical for correctly framing the provisions of such Acts. The lack of rigour is on two fronts: first, in obtaining adequate scientific knowledge related to resource dynamics, use behaviour and their impacts; and second, in understanding the physical, socio-economic and institutional contexts in which these legislations must work.

2. Patchy Water Laws

Most often, it is argued that water-related Acts in India fail because they lack legal teeth. This argument is often made in the context of the Water Pollution (Prevention and Control) Act of 1974. What is basically meant by this statement is that there are no provisions in the Act for the agencies tasked with implementing the Act to penalize violators. However, this is far from the truth. The existing provisions in the law enable the enforcing agencies to take the violators to trial in the normal courts of law. This has been done in many cases of pollution in India, wherein cases were filed against the polluting industries by the state or central pollution control board (as the case may be) as Public Interest Litigations (PILs), and industries were ordered by the courts to pay compensation to the affected parties for the damages caused by them. There are many such cases in India today – starting from the Fatehwadi canal command area, which receives water from a barrage on the Sabarmati River in Ahmedabad, Gujarat, to the Noyyal River in Coimbatore, Tamil Nadu, to Pali in Rajasthan.

Contrary to this widespread notion, the problem with most water-related Acts that render them ineffective or even redundant is that the provisions of those Acts are very patchy (Panickar, 2007), incoherent and inconsistent, with many gaps and loopholes that make them susceptible to different interpretations. One hypothetical example is penalizing an industry for polluting a water body because it disposes of untreated effluents into it. Unless it is clearly specified in the Act or in the rules that are framed thereafter 'what the maximum allowable limit of pollutant concentration in the effluent for discharge into the water body is', it is difficult to judge whether the particular industry is culpable for pollution of the water body or not. It is also possible that the maximum pollutant concentration or desirable quality of water in the receiving water body is prescribed (in terms of BOD, COD, TDS, dissolved oxygen or pH) and not that of the potential polluting agents, which is the case at present in the Pollution (Control & Prevention) Act of 1974. Such definitions will also not be adequate to penalize a particular polluting agent if there are many polluting sources that result in the pollution concentration exceeding certain limits.

This is because in such situations, it is difficult to attribute the pollution (or water quality parameters exceeding certain permissible limits) to one agent, which makes conviction difficult.

This is contrary to what has been seen in the case of the water acts of many developed countries (countries of the European Union, Canada, the United States and Japan) that were reviewed and presented in Chapter 2 (of this monograph). Those Acts cover many thematic areas (such as water supply service levels with regard to quantity and quality; parameters to be considered for checking the potability of the water supplied and their permissible values; source protection and penalties for polluting source water; standards for the facilities to be created for water supply and wastewater treatment; duties and powers of the water service agencies) and are quite specific. For instance, the Drinking Water Regulations (2014) of the European Union set the standards for the analysis of water quality parameters in laboratories.

Another issue with certain Acts in the water sector is that they can be easily contested in a court of law because of the rationale on which they are built. One example is the erstwhile Groundwater (Protection of Drinking Water Sources) Act of Maharashtra (GOM, 1993), meant for protecting drinking water sources, which says that no irrigation well shall be drilled within a distance of 500 m from a public water source (PWS) that taps groundwater. Such provisions are innocuous for two reasons: First, it is extremely difficult to say exactly what should be the minimum distance between wells beyond which well interference cannot happen, as 'well interference' is a function of many complex factors, including the characteristics of the aquifer (specific yield and transmissivity) and the pumping rates. Unless and until the enforcing agency can prove in a court of law that the irrigation well is affecting the performance of the drinking water well in terms of yield reduction, etc., the reported 'offence' cannot be established, and the plaintiff can get relief under the 'right to life', which is a fundamental right enshrined in Article 21 of the Constitution of India that can take precedence over all such Acts (source: based on Upadhyay, 2011)[1]. Therefore, the statements of the provisions of the Act need to be qualified with further details of the geological formation and the rate of

1 The right to 'pollution-free water' and the right of access to 'safe drinking water' have been read as part of 'Right to Life' under Article 21 of the Constitution of

pumping for the irrigation well. Secondly, as pointed out by Phansalkar and Kher (2006), the Act does not bring under its ambit the myriad irrigation wells that are located beyond the 500 m radius of the PWS but tap water from the same aquifer that the PWS taps and therefore can threaten its sustainability. Such provisions make the Act redundant.

3. Laws without a Powerful Agency to Enforce

This is perhaps the most common situation encountered in India. In most instances, it is clear in the Act which agency would be responsible for implementing or enforcing it. However, the administrative capacity of the agency to enforce the Act is questionable. This is quite evident in the case of groundwater legislation. In the states where groundwater legislation is enacted, the agency responsible for investigation and planning is mandated to monitor groundwater use and determine if any violations of the Act have taken place. However, these agencies have relatively very thin staffs compared to the magnitude of work involved in monitoring activities such as well drilling and well deepening, most of which occur in rural areas, and then checking whether such activities are in compliance with or in violation of the provisions of the Act. It is quite unlikely that such agencies would deploy their staff to monitor such activities *suo moto*, given the resource constraints, and it is also very unlikely that the villagers would voluntarily report such incidents to the agency staff. The reason for the latter is that the people who would probably violate such Acts in the future would be very large in number, and no one would like to annoy fellow villagers by reporting an incident like this. Under such circumstances, the success of implementing an Act like the one mentioned above would depend entirely on the agency staff's own capacity, motivation and commitment.

India. This is a judicially evolved fundamental right to water, and such a right is not explicitly incorporated under the Constitution (Upadhyay, 2011).

The same is also the case with the Maharashtra Water Resources Regulatory Authority Act, 2005 (*as modified up to 23 January 2018*) (GoM, 2018). The Act provides for allocating (surface) water in different river basins of the state amongst different competitive use sectors and also among users within the sectors, based on well-defined volumetric 'water-entitlements' and pre-determined prices for each sectoral use. But the Maharashtra Water Resources Regulatory Authority (MWRRA), which is mandated to enforce the Act, is a small, centralized body with thin staff, having no local arms to monitor water use by the millions of individual water users (farmers, drinking water users, industries, etc.). While these agencies mentioned above can hire additional staff and deploy them for monitoring the activities in the field and detect any violations of the Act, this can immensely increase the transaction costs, thereby bringing down the net benefits that society can accrue from enforcing the law.

4. 'Disconnect' between the Provisions of the Law and the Realities on the Ground

Several Acts do not reflect the situation on the ground and appear very utopian. One example is the Act that makes rooftop water harvesting mandatory for households in urban areas, like the ordinance passed by the government of Tamil Nadu in 2003, which is known as the 'Tamil Nadu Municipal Laws (Second Amendment) Ordinance, 2003' (GoTN, 2003b). As per the ordinance, it is mandatory for every urban housing stock to have a rooftop rainwater harvesting system on its premises. The ordinance is completely silent about the fact that a significantly large proportion of the people in urban areas, especially those living in slums and unauthorized colonies, lack proper roofs with sufficient area, and for them, building a rooftop water harvesting system is of no use. In such situations, most of them will not even have the space to construct the tanks. Therefore, it is important that those who frame such Acts keep these socio-economic realities in mind and that the Act explicitly states who would come under its purview.

Metering of water supply and volumetric pricing are part of the reform agenda in the urban water sector of India (Mathur and Thakur, 2003; Lee *et al.*, 2016). Therefore, a law that can be conceived by urban water utilities is that of making metering of domestic water consumption in urban areas mandatory for every household, affecting pro rata (consumption-based) pricing of water. Here also, the Act can exclude a large number of households without realizing the intended outcomes, the reason being that a very large proportion of the people in urban areas live in housing stocks that have a single 'municipal water connection' obtained in the name of the housing society they are in, and many living in slums do not have individual water connections. The housing society in turn distributes the water from a sump among the households using its own mini distribution system, often making the water supply round-the-clock (24 × 7). Such a mini water supply system helps avoid the need for individual storage tanks for each household that are otherwise required to deal with problems arising out of intermittent water supply from the municipality or the corporation. In such situations, it is important that the housing societies frame their own rules and make arrangements for metering water consumption by individual households for the Act to produce the desired effects. Unless this happens, the intended outcomes of water pricing by the municipality will not be realized. Similarly, the intended impacts of pro-rata pricing of water will not be achieved if there are too many people living in slums without individual water connections and who are dependent on common standposts.

5. Water Laws Are Different from Other Laws

Most water-related laws fail to recognize that water is a spatially and temporally dynamic resource. The resource condition, in terms of both quantity and quality, varies with space and time. In a country like India, where most precipitation occurs during the monsoon season, the condition of water during the monsoon is very different from that during the summer season almost everywhere in the country. Such changes in resource conditions alter the requirements at the grassroots level, or, in other words, what is desirable for the larger societal good. The quantum of pollutants that a river can assimilate during the wet season is much higher than during the dry season when the flows are extremely limited. Therefore, the level of treatment of the effluent that enters the river must ensure that the quality (physical, chemical and biological) of the river water does not exceed permissible levels even during the driest part of summer.

Similarly, the extent of pumping that an aquifer can support during the monsoon season is much higher than that of the summer season, and in many situations, a reduced level of pumping during the monsoon season cannot result in increased water availability during the summer season due to the non-availability of storage space in the aquifer (in hard rock areas) and the lateral flows. Therefore, it would make sense to maintain a high level of abstraction during the wet season. However, the Acts generally do not take cognizance of such complex situations and are often too broad, generally covering the entire geographical area of a state. But the conditions with respect to natural water endowment (surface runoff in rivers, groundwater replenishment, soil moisture conditions, etc.) and water demands can change drastically between regions within a state, and Acts are often not capable of handling the situations arising out of such conditions. Essentially, it means that some regions would require regulations on water use, including effluent discharge, whereas some other regions may not require any such stringent regulations. This basically means that the larger state-wide support for such laws is unlikely to emerge from the general public.

The complex resource dynamics also make it difficult to apply the provisions of certain Acts meaningfully. Groundwater and surface water (lakes, streams and rivers) are part of the same hydrological system and often interact (Winter *et al.*, 1998). But different legal principles are applied to regulate the development and use of these two resources. While all rights over surface water resources (to appropriate, allocate and prevent their contamination) are entrusted to the state, their interconnectedness with groundwater systems makes it necessary for the law governing the use of surface water to also apply to groundwater wherever such hydraulic interconnectedness exists (Cullet, 2012). Since it might be very difficult to establish such phenomena in a court of law without proper scientific studies to produce the required evidence, the present legal framework might end up giving the landowners overbearing power over groundwater.

Research studies have shown that excessive groundwater draft in the upper catchment caused reductions in lean season flows in large river basins (Kumar (2010) in the case of the Narmada River basin) and inflows into tanks (Kumar and Vedantam (2016) in the case of tanks in erstwhile Andhra Pradesh) through reductions in natural discharge from aquifers to the streams. However, the state governments cannot act against such users precisely because such uses are not apparent. An isotope study carried out in the Upper Jhelum basin in Kashmir indicated that the shrinking of recharge areas (due to construction of roads along stream banks, conversion of small ponds into cropland, etc.) and unplanned pumping were the major factors responsible for the drying up of the springs and wells in the alluvial aquifers (Jeelani *et al.*, 2021). Thambe et al. (2012) reported a 48 per cent reduction in the spring discharge in drought-prone areas and 35 per cent in other areas of the Sikkim Himalaya.

As one can infer, the performance and overall sustainability of drinking water sources that tap water directly from springs, streams, rivers and lakes will be heavily dependent on the lean season flows, which can be severely altered by excessive pumping of groundwater from the shallow aquifers that are hydraulically linked to these surface water bodies. Hence, this is an important concern that has to be dealt with through comprehensive legislation.

6. Laws Being Framed without Taking Cognizance of the Socio-economic Consequences

Many laws fail in implementation just because they are not socially viable or have serious conceptual flaws. One example is a law that bans the drilling of new tube wells in areas classified as 'over-exploited' or 'dark' by the groundwater agency (the state groundwater department or the Central Ground Water Board). The law, if effectively implemented, will completely deprive non-well-owning farmers of directly accessing irrigation water from the common pool. What is often ignored by those who draft the law is that the livelihood of the farmer comes under threat with the implementation of such a law. Given the fact that in such areas, several thousands of farmers might have already been using deep tube wells for irrigation for decades before the aquifer became 'over-exploited', the issue has intra-generational equity dimensions. What such a law essentially does is stop a new entrant (most likely a less resource-rich farmer), who probably might have been dependent on water purchases for a long time, from owning a well of his/her own to improve his/her livelihoods, while the resource-rich continue tapping the over-exploited aquifer. This is against the theory of natural justice. The issue here is one of 'intra-generational equity'. What might actually happen on the ground is a sudden increase in 'rent-seeking behaviour' among the officials who are mandated to monitor the activities on the ground, wherein they might start asking for bribes from the farmers who want to drill wells, flouting the law.

It is also true that in certain cases, the benefits of implementing an Act are indirect (largely through positive externalities). Examples are: (i) the public health benefits through the implementation of a local pollution control law, resulting from reduced pollution of a river/stream or cleaning of a lake that is directly used by communities for domestic purposes, in the form of reduced incidence of waterborne diseases (Kumar and Tortajada, 2020); and (ii) the direct benefits to the water supply utilities in the form of lower cost of treatment of water prior to supplying it to the consumers, resulting from reduced pollution of natural water bodies (Dearmont *et al.*, 1998)[2]. Not all of these benefits are tangible and immediate. Therefore, while introducing a pollution control act in a locality, educating the communities and raising their awareness levels about the public health impacts of

2 Dearmont *et al.* (1998) found in a case study in Texas City that chemical contamination of surface water increased the cost of municipal water treatment by US $25 per million litres of water.

water pollution will be important for them to foresee the private benefits of implementing the Act in the long run and value them in monetary terms, thereby increasing their willingness to support such an Act. Therefore, for laws that have negative socioeconomic consequences to succeed, it is important that their indirect benefits, which are often not perceived, are highlighted for them to gain social legitimacy.

7. Summary and Conclusions

From the foregoing discussions, it is clear that for any legislation to succeed, it has to have a scientific rationale that its implementation would actually produce the intended benefits to society at large. The point is that improper implementation of the provisions of the legislation can produce unintended consequences for society (Tyagi, 2021). Building such a rationale would require a thorough analysis of the physical, socioeconomic and institutional environments in which the proposed legislation will be embedded. The legislation should be clear and unambiguous and should not be liable to different interpretations (Schane, undated) because, as Mellinkoff (1963) pointed out, 'The Law is a profession of words', and the language used matters. Further, the offences as per the various provisions of the law should be provable in a court of law through scientific evidence.

Furthermore, no provision of the law should be in violation of the fundamental rights of the citizens enshrined in the Constitution for it to be legally tenable, such as the 'right to life'. The implementation of the law should produce a significant benefit to society, which, when weighed against the costs that society has to incur, should be much greater. However, some of the benefits are indirect (in the form of positive externalities), and therefore educating and raising the level of awareness of the community in order for them to visualize those intangible benefits and attach a value to them will be crucial. The transaction cost of enforcing the law or the Act on the ground in letter and spirit should be an important consideration while evaluating the benefits. A piece of legislation that adversely affects the different segments of the same community or group differentially is likely to face resistance and opposition from those segments that are worst-affected. Finally, what the legislation mandates should reflect the realities on the ground for most people to follow. Obviously, a lot has to go into drafting such Acts that meet these basic requirements for them to be meaningful.

Bibliography

Agarwal, A., Narain, S. (1997) *Dying Wisdom: Rise and Fall of Traditional Water Harvesting Systems.* New Delhi: Centre for Science and Environment.

Alaerts, G. J., Blair, T. L. and Hartvelt, F. J. A. (eds) (1991) *A Strategy for Water Sector Capacity Building: Proceedings of the UNDP Symposium, Delft, 3–5 June, 1991.* Delft, The Netherlands: IHE Report Series 24, pp. 185–191.

Arabindoo, P. (2011) 'Mobilising for Water: Hydro-politics of Rainwater Harvesting in Chennai', *International Journal of Urban Sustainable Development*, 3(1), pp. 106–126. Available at: <https://doi.org/10.1080/19463138.2011.582290>.

Asian Development Bank. (2007) *2007 Benchmarking and Data Book of Water Utilities in India.* A partnership between the Ministry of Urban Development, Government of India and the Asian Development Bank.

Bassi, N. and Kumar, M. D. (2012) 'Addressing the Civic Challenges', *Environment and Urbanization Asia*, 3(1), pp. 165–183. Available at: <https://doi.org/10.1177/097542531200300109>.

Bhaduri, A. (2015) *Bisalpur Revisited – 10 Years after Protesters Were Shot, Killing 5*, India Water Portal, 10 September 2010. Available at: <https://www.indiawaterportal.org/articles/bisalpur-revisited-10-years-after-protesters-were-shot-killing-5>.

Birkenholtz, T. (2010) '"Full-Cost Recovery": Producing Differentiated Water Collection Practices and Responses to Centralized Water Networks in Jaipur, India', *Environment and Planning A*, 42(9), pp. 2238–2253. Available at: <https://doi.org/10.1068/a4366>.

Boyd, J. (2000) *The New Face of the Clean Water Act: A Critical Review of the EPA's Proposed TMDL Rules.* Washington, DC: Resources for the Future. Available from <http://www.rff.org/Documents/RFF-DP-00-12.pdf>.

Brikké, F. and Bredero, M. (2003) *Linking Technology Choice with Operation and Maintenance in the Context of Community Water Supply and Sanitation: A Reference Document for Planners and Project Staff.* Geneva, Switzerland: World Health Organization and IRC Water and Sanitation Centre.

Calder, I. R. *et al.* (2008) 'Watershed Development in India. 2. New Approaches for Managing Externalities and Meeting Sustainability Requirements', *Environment, Development and Sustainability*, 10(4), pp. 427–440. Available at: <https://doi.org/10.1007/s10668-006-9073-0>.

Carter, R. C., Tyrrel, S. F., Howsam, P. (1997) 'The impact and sustainability ofwater and sanitation programmes in developing countries', *Journal of the Chartered Institution of Water and Environmental Management*, 13, pp. 292–296.

Central Pollution Control Board (CPCB). (2006a) *Guidelines for Water Quality Management*. Delhi: CPCB.

Central Pollution Control Board (CPCB). (2006b) *Status of Sewage Treatment in India*. Delhi: CPCB.

Central Pollution Control Board. (2008) *Evaluation of Operation and Maintenance of Sewage Treatment Plants in India-2007*. New Delhi: Central Pollution Control Board, Ministry of Environment and Forests.

Central Water Commission. (2010) *Pricing of Water in Public System in India*. Delhi: Information Systems Organisation, Water Planning & Projects Wing, Central Water Commission.

Chaudhuri, S. *et al.* (2020) 'Water for All (*Har Ghar Jal*): Rural Water Supply Services (RWSS) in India (2013–2018), Challenges and Opportunities', *International Journal of Rural Management*, 16(2), pp. 254–284. Available at: <https://doi.org/10.1177/0973005220946661>.

Cullet, P. (2007) Water Law in India: Overview of Existing Frame Work and Proposed Reforms. IELRC working paper 2007–01, Geneva. Available at: <https://www.ielrc.org/content/w0701.pdf>.

Cullet, P. (2012) 'Groundwater: Towards a New Legal and Institutional Framework', in P. Cullet, S. Paranjape, H. Thakkar, M. S. Vani, K. J. Joy, and M. K. Ramesh (eds), *Water Conflicts in India: Towards a New Legal and Institutional Framework*. Pune: Forum for Policy Dialogue on Water Conflicts in India.

Cullet, P. (2016) 'Governing Water to Foster Equity and Conservation: Need for New Legal Instruments', *Economic and Political Weekly*, 51(53). (31 December 2016), available at http://www.epw.in/node/148151/pdf.

Cullet, P. and Gupta, J. (2009) 'India: Evolution of Water Law and Policy', in Joseph W. Dellapenna and Joyeeta Gupta (eds), *The Evolution of the Law and Politics of Water*. Dordrecht: Springer Academic Publishers, p. 159.

Dearmont, D., McCarl, B. A. and Tolman, D. A. (1998) 'Costs of Water Treatment Due to Diminished Water Quality: A Case Study in Texas', *Water Resources Research*, 34(4), pp. 849–853. <https://doi.org/10.1029/98wr00213>.

Gleick, P. (1996) 'Basic Water Requirements for Human Activities: Meeting Basic Needs', *Water International*, 21, pp. 83–92.

Gleick, P. (ed.). (1993) *Water in Crisis: A Guide to the World's Freshwater Resources*. New York: Oxford University Press.

Howard, G. (2001) Challenges in increasing access to safe water in urban Uganda: economic, social and technical issues, in G. F. Craun, F. S. Huachman, and D. E. Robinson (eds), *Safety of Water Disinfection: Balancing Microbial and Chemical Risks*.Washington, D. C.: ILSI Publications, pp. 483–499.

Devi, P. R. (1991) 'Groundwater Development and Legal Regulation', *Journal of the Indian Law Institute*, 33(4), pp. 614–634.

The Drinking Water Protection Act, SBC 2001, c. 9 (Canada). Available at: <https://canlii.ca/t/5316z> (Accessed: 25 January 2022).

DRWH Water Quality: A Literature Review Padma Vasudevan and Namrata Pathak IIT Delhi European Commission, Mile stone1: Reports.

European Union (Drinking Water) Regulations 2014 (S.I. No. 122 of 2014). Dublin: Stationery Office. Available at: <https://www.irishstatutebook.ie/eli/2014/si/122/made/en/pdf>

Foundation for Water Research. (2004) *Payment Strategies and Price Elasticity of Demand for Water for Different Income Groups in Three Selected Urban Areas*. Available at: <http://www.fwr.org/wrcsa/1296104.htm>.

Ghana Water and Sewerage Corporation Act 1965 *(Act 310)*, Republic of Ghana.

Global Water Partnership. (2000) *Towards Water Security: A Framework for Action*. Stockholm, Sweden: GWP.

Global Water Partnership. (2017) *Elements of Water Law (A2.01), Global Water Partnership*. Available at: <https://www.gwp.org/en/learn/iwrm-toolbox/The-Enabling-Environment/Legal-Framework/Elements-of-Water-Law/> (Accessed: 6 February 2022).

Government of Andhra Pradesh. (2002) *Andhra Pradesh – An Act to Promote Water Conservation, and Tree Cover and Regulate the Exploitation and Use of Ground and Surface Water for Protection and Conservation of Water Sources, Land and Environment and Matters, Connected Therewith or Incidental Thereto, Act. No. 10 of 2022*. Hyderabad: Government of Andhra Pradesh. Available at: <http://forests.telangana.gov.in/Documents/FCA/WALTA%20Act%202002.pdf>

Government of India. (2012) *Groundwater Year Book 2011–12*. Faridabad: Central Groundwater Board, Ministry of Water Resources, Government of India.

Government of Karnataka. (2011) *Karnataka Groundwater (Regulation and Control of Development and Management) Act 2011*. Bangalore: Government of Karnataka.

Government of Maharashtra. (1993) *Maharashtra Groundwater (Regulation for Drinking Purpose) Act 1993*. Mumbai: Government of Maharashtra.

Government of Maharashtra. (2013) *Maharashtra Groundwater (Development and Management) Act 2013 (MAHARASHTRA ACT No. XXVI OF 2013)*. Mumbai: Government of Maharashtra.

Government of Maharashtra. (2018) *Maharashtra Water Resources Regulatory Authority Act, 2005 (as Modified Up to 23 January 2018), MAHARASHTRA*

ACT No. XVIII OF 2005. Mumbai: Law and Judiciary Department, Government of Maharashtra.

Government of Rajasthan. (2010) *State Water Policy*. Jaipur: State Water Resource Planning Department, Government of Rajasthan.

Government of Tamil Nadu. (2003a) *The Tamil Nadu Groundwater (Development and Management) Act 2003 (Tamil Nadu Act 3 of 2003)*. Chennai: Government of Tamil Nadu.

Government of Tamil Nadu. (2003b) *Tamil Nadu Ordinance No. 4 of 2003*. Chennai: Tamil Nadu Government Gazette Extraordinary.

Grieve, C. and Johnson-Jones, G. (2000) 'Management of Unaccounted for Water in India', *Proceedings of 26th WEDC Conference on Water, Sanitation and Hygiene: Challenges of the Millennium*, Dhaka, Bangladesh.

The Ground Water Protection Regulation, BC Reg 299/2004 (Canada). Available at: <https://canlii.ca/t/jwd6> (Accessed: 25 January 2022).

The Health Hazards Regulation, BC Reg 216/2011 (Canada). Available at: <https://canlii.ca/t/54p9v> (Accessed: 25 January 2022).

Helmer, R. G. and Hespanhol, I. (eds). (1997) *Water Pollution Control: A Guide to the Use of Water Quality Management Principles*. London: E & FN Spon.

Hemani, R., Kumar, M. D., Bassi, N. and Chandra, U. (2021) 'Mapping Climate-Induced Risk for Water Supply, Sanitation and Hygiene in Rajasthan', in M. D. Kumar, Y. Kabir, R. Hemani and N. Bassi (eds), *Management of Irrigation and Water Supply under Climatic Extremes: Empirical Analysis and Policy Lessons from India*. Switzerland: Springer, pp. 241–286.

Hodgson, S. (2004) *Land and Water – the Rights Interface (FAO Legislative Study 84, for the Development Law Service)*. Rome: Food and Agriculture Organization of the United Nations.

Hooda, S. M. (2013) *India – Rajasthan Water Assessment and Potential for Private Sector Interventions*. Washington, DC: World Bank Group. Available at: <http://documents.worldbank.org/curated/en/261801490687132806/India-Rajasthan-water-assessment-and-potential-for-private-sector-interventions> (Accessed: 17 February 2023)

Howard, G. (2002) *Water Supply Surveillance: A Reference Manual*. England: WEDC, Loughborough University.

Hutton, G. and Chase, C. (2017) 'Water Supply, Sanitation and Hygiene', in C. N. Mock, R. Nugent, O. Kobusingye and K. R. Smith (eds), *Injury Prevention and Environmental Health*, 3rd edn. Washington, DC: The International Bank for Reconstruction and Development/The World Bank.

International Environmental Law Research Centre. (2007) *Draft Guidelines for Preparation of Legislation for Framing Drinking Water Regulations, 2007*. Available at: <www.ielrc.org/content/e0714.pdf>.

International Environmental Law Research Centre. (no date) *India – State-Level Water Law Instruments.* Available at: <http://www.ielrc.org/water/doc_states.php>.

Jat, M. L., Jain, P. M., Sharma, S. K. and Jain, L. K. (2009) 'Water Resources Management in a Water-Deficit State', *Journal of Indian Water Resources Society*, 29(3), pp. 1–7.

Jeelani, G., Lone, S. A., Lone, A. and Deshpande, R. D. (2021) 'Groundwater Resource Protection and Spring Restoration in Upper Jhelum Basin (UJB), Western Himalayas', *Groundwater for Sustainable Development*, 15, p. 100685. <https://doi.org/10.1016/j.gsd.2021.100685>.

Kumar, M. D. (2004) 'Roof Water Harvesting for Domestic Water Security: Who Gains and Who Loses?', *Water International*, 29(1), pp. 43–53.

Kumar, M. D. (2006) *Water Management in River Basins: A Study of Sabarmati River Basin in Gujarat*, PhD thesis, Faculty of Management, Sardar Patel University, VV Nagar, Gujarat.

Kumar, M. D. (2009) *Water Management in India: What Works, What Doesn't.* New Delhi: Gyan Publishing House.

Kumar, M. D. (2010) *Managing Water in River Basins: Hydrology, Economics, and Institutions.* New Delhi: Oxford University Press.

Kumar, M. D. (2014) *Thirsty Cities: How Indian Cities Can Meet Their Water Needs.* New Delhi: Oxford University Press.

Kumar, M. D. (2018) *Water Policy, Science and Politics: An Indian Perspective.* Amsterdam: Elsevier Science.

Kumar, M. D. and Bassi, N. (2013) *Water Accounting Study of Luni River Basin and District IWRM Planning for Pali*, Final Report Submitted to the Technical Assistance Team of the EUSPP, Jaipur, Rajasthan, December 2013.

Kumar, M. D., Bassi, N. and Kumar, S. (2022) *Drinking Water Security in Rural India: Dynamics, Influencing Factors, and Improvement Strategy.* Singapore: Springer.

Kumar, M. D., Bassi, N., Venkatachalam, L., Sivamohan, M. V. K. and Niranjan, V. (2012) *Capacity Building in Water Resources Sector of India.* IRAP Occasional Paper No. 5. Hyderabad: Institute for Resource Analysis and Policy.

Kumar, M. D., Ghosh, S., Patel, A., Singh, O. P. and Ravindranath, R. (2006) 'Rainwater Harvesting in India: Some Critical Issues for Basin Planning and Research', *Land Use and Water Resources Research*, 6, pp. 1–17. <https://doi.org/10.22004/ag.econ.47964>.

Kumar, M. D., and Pandit, C. M. (2020) India's water management debate: is the 'civil society' making it everlasting?, in Kumar, M. D. (ed.), *Politics and Policies for Water Resources Management in India.* Singapore: Taylor and Francies, pp 28–41.

Kumar, M. D., Patel, A., Ravindranath, R., Singh, O. P. (2008) 'Chasing a Mirage: Water Harvesting and Artificial Recharge in Naturlaly Water-Scarce Regions', *Economic and Political Weekly*, 43(35), pp. 61–71.

Kumar, M. D. and Tortajada, C. (2020) *Assessing Wastewater Management in India, Springer Briefs in Water Science and Technology*. Singapore: Springer.

Kumar, M. D. and Vedantam, N. (2016) 'Groundwater Use and Decline in Tank Irrigation? Analysis from Erstwhile Andhra Pradesh', in M. D. Kumar, A. J. James and Y. Kabir (eds), *Rural Water Systems for Multiple Uses and Livelihood Security*. Amsterdam: Elsevier Science, pp. 145–182.

Lee, F., Aggarwal, V. and Nickum, J. (2016) 'Urban Domestic Water Pricing in India and China', *Water Policy*, 18(S1), pp. 68–82. <https://doi.org/10.2166/wp.2016.205>.

Mathur, O. P. and Thakur, S. (2003) *Urban Water Pricing: Setting the Stage for Reforms*. New Delhi: National Institute of Urban Affairs.

Mellinkoff, D. (1963) *The Language of the Law*. Boston: Little, Brown & Co.

Mittal, A. K. (2004) 'Managerial and Financial Aspects of Sustainable WATSAN Operation: A Case of Rajasthan, India', in *Conference on Market Development of Water & Waste Technologies through Environmental Economics*, Paris, 28–29 May.

The Municipal Wastewater Regulation, BC Reg 87/2012 (Canada). Available at: <https://canlii.ca/t/5377c> (Accessed: 25 January 2022).

National Disaster Management Authority. (2010) *Plan to Counter the Threats to Municipal Water Supply and Water Reservoirs*. Delhi: National Disaster Management Authority.

National Rural Drinking Water Programme. (2013) *Movement towards Ensuring People's Drinking Water Security in Rural Areas-Framework for Implementation (Updated 2013)*. New Delhi: National Rural Drinking Water Programme, Rajiv Gandhi Rural Drinking Water Mission, Ministry of Drinking Water, Govt. of India.

National Water Policy. (2012) *National Water Policy 2012*. New Delhi: Ministry of Water Resources, Government of India.

Noll, R., Shirley, M. M. and Cowan, S. (2000) 'Reforming Urban Water Systems in Developing Countries', in Anne O. Krueger (ed.), *Economic Policy Reform: The Second Stage*. Chicago, IL: University of Chicago Press, pp. 243–288.

Olmstead, S., Hanemann, W. M. and Stavins, R. (2007) 'Water Demand under Alternative Price Structures', *Journal of Environmental Economics and Management*, 54(2), pp. 181–198. <https://doi.org/10.3386/w13573>.

Padhi, B. K., Baker, K. K., Dutta, A., Cumming, O., Freeman, M. C., Satpathy, R., Das, B. S. and Panigrahi, P. (2015) 'Risk of Adverse Pregnancy Outcomes among Women Practicing Poor Sanitation in Rural India: A Population-Based

Prospective Cohort Study', *PLoS Medicine*, 12(7). <https://doi.org/10.1371/journal.pmed.1001851>.

Padma, V. and Namrata, P. (2001) *DRWH Water Quality: A Literature Review*. Delhi: Centre for Rural Development and Technology, Indian Institute of Technology.

Panickar, M. (2007) *State Responsibility in the Drinking Water Sector: An Overview of the Indian Scenario*. IELRC Working Paper # 06. New Delhi.

Phansalkar, S. and Kher, V. (2006) 'A Decade of the Maharashtra Groundwater Legislation: Analysis of the Implementation Process', *Law, Environment and Development Journal*, 2(1), pp. 67–83.

Pisharoty, P. R. (1990) *Characteristics of Indian Rainfall*. Ahmedabad: Monograph, Physical Research Laboratories.

Rajaram, T. and Das, A. (2008) 'Water Pollution by Industrial Effluents in India: Discharge Scenarios and Case for Participatory Ecosystem Specific Local Regulation', *Futures*, 40(1), pp. 56–69. <https://doi.org/10.1016/j.futures.2007.06.002>.

Ray, S. and Bijarnia, M. (2006) 'Upstream vs Downstream: Groundwater Management and Rainwater Harvesting', *Economic and Political Weekly*, 41(23), pp. 2375–2383.

Reddy, V. R., Jayakumar, N., Venkataswamy, M., Snehalatha, M. and Batchelor, C. (2012) 'Life-Cycle Costs Approach (LCCA) for Sustainable Water Service Delivery: A Study in Rural Andhra Pradesh, India', *Journal of Water Sanitation and Hygiene for Development*, 2(4), pp. 279–290. <https://doi.org/10.2166/washdev.2012.062>.

Reddy, V. R., Rao, M. S. R. and Venkataswamy, M. (2010) *'Slippage': The Bane of Drinking Water and Sanitation Sector (A Study of Extent and Causes in Rural Andhra Pradesh)*. WASH Cost India-CESS Working Paper No. 87. Hyderabad: Centre for Economic and Social Studies.

Renwick, M. E., Green, R. D. and McCorkle, C. (1998) *Measuring the Price Responsiveness of Residential Water Demand in California's Urban Areas*. A Report Prepared for the California Department of Water Resources.

Safe Drinking Water Act 1974–2004 (USA).

Schane, S. (undated) *Ambiguity and Misunderstanding in the Law*. Available at: <https://idiom.ucsd.edu/~schane/law/ambiguity.pdf> (Accessed: 26 January 2022).

Singh, A. (2012) '"Damned" Ramgarh Still Dry', *The Times of India*, 24 August. Available at: <https://timesofindia.indiatimes.com/city/jaipur/damned-ramgarh-still-dry/articleshow/15626741.cms> (Accessed: 25 January 2022).

South Eleuthera Water Supply Act 1962 (The Bahamas).

SPHERE. (1998) *The Sphere Project: Humanitarian Charter and Minimum Standards in Disaster Response*. Geneva, Switzerland.

State Water Investigation Directorate. (2006) *West Bengal Ground Water Resources (Management, Control & Regulation) Act*. Kolkata: State Water Investigation Directorate.

State Water Resources Planning Department. (2013) *Annual Report 2013 of State Water Resources Planning Department on Environmental Status*. Jaipur: SWRPD.

Thambe, S., Kharel, G., Arrawatia, M. L., Kulkarni, H., Mahamuni, K. and Ganeriwala, A. K. (2012) 'Reviving Dying Springs: Climate Change Adaption Experiments from Sikkim Himalaya', *Mountain Research and Development*, 32(1), pp. 62–72.

Tyagi, B. (2021) 'Good Laws, Bad Implementation: Biggest Problem in India', *The Daily Guardian*, 17 February 2021. Available at: <https://thedailyguardian.com/good-laws-bad-implementation-biggest-problem-in-india/#:~:text=The%20problem%20is%20not%20only,strict%20laws%20for%20these%20matters> (Accessed: 26 January 2022).

UN-Habitat. (2012) *Reduction of Illegal Water, Utility Management Series for Small Towns*, Vol. 6. Nairobi, Kenya: UN-Habitat.

United Nations Development Programme. (2006) *Human Development Report 2006: Beyond Scarcity: Power, Poverty and the Global Water Crisis*. New York: UNDP.

Upadhyay, V. (2011) *Water Rights and 'New' Water Laws in India: Emerging Issues and Concerns in a Rights Based Perspective, India Infrastructure Report 2011*. Available at: <https://www.idfc.com/pdf/report/2011/Chp-5-Water-Rights-And-The-New-Water-Laws-In-India.pdf> (Accessed: 25 January 2022).

Wagner, E. G. and Lanoix, J. N. (1959) *Water Supply for Rural Areas and Small Communities*. Geneva, Switzerland: World Health Organization.

Water Services Act 1997 (Republic of South Africa).

Water Services Act 2007 (Ireland). Dublin: Stationery Office. Available at: <http://www.irishstatutebook.ie/pdf/2007/en.act.2007.0030.pdf> (Accessed: 12 February 2022).

Water Supply (Federal Territory of Kuala Lumpur) Act 1998 (Malaysia). The Commissioner of Law Revision.

Water Supply Regulations 2010, Water, England and Wales.

Waterworks Act 1957 (Japan).

Winter, T. C., Harvey, J. W., Franke, O. L. and Alley, W. M. (1998) 'Ground Water and Surface Water: A Single Resource', *U.S. Geological Survey Circular*, 1139. <https://doi.org/10.3133/cir1139>.

World Bank Analytical and Advisory Assistance (AAA) Program China: Addressing Water Scarcity – From Analysis to Action Policy Note (December, 2007),

Water Supply Pricing in China: Economic Efficiency, Environment, and Social Affordability.

World Health Organization. (1997) *Guidelines for Drinking-Water Quality*, 2nd edn. *Volume 3: Surveillance and Control of Community Supplies.* Geneva: World Health Organization.

World Health Organization and UNICEF. (2000) Global Water Supply and Sanitation Assessment 2000, Report. World Health Organization/United Nations Children's Fund, Geneva/New York.

World Health Organization. (2008) *Guidelines for Drinking-Water Quality*, 3rd edn. *Volume 1: Recommendations Incorporating the First and Second Addenda.* Geneva: World Health Organization.

World Health Organization. (2011) *Guidelines for Drinking-Water Quality: Surveillance and Control of Community Supplies*, 2nd edn, Vol. 3. Geneva: World Health Organization.

WWAP (World Water Assessment Programme). (2012) *The United Nations World Water Development Report 4: Managing Water under Uncertainty and Risk.* Paris: UNESCO.

Index

www.ingramcontent.com/pod-product-compliance
Ingram Content Group UK Ltd.
Pitfield, Milton Keynes, MK11 3LW, UK
UKHW021012120926
8556IPUK00039B/1035